Lecture Notes in Electrical Engineering

Volume 502

**** Indexing: The books of this series are submitted to ISI Proceedings, EI-Compendex, SCOPUS, MetaPress, Springerlink ****

Lecture Notes in Electrical Engineering (LNEE) is a book series which reports the latest research and developments in Electrical Engineering, namely:

- Communication, Networks, and Information Theory
- Computer Engineering
- Signal, Image, Speech and Information Processing
- Circuits and Systems
- Bioengineering
- Engineering

The audience for the books in LNEE consists of advanced level students, researchers, and industry professionals working at the forefront of their fields. Much like Springer's other Lecture Notes series, LNEE will be distributed through Springer's print and electronic publishing channels.

For general information about this series, comments or suggestions, please use the contact address under "service for this series".

To submit a proposal or request further information, please contact the appropriate Springer Publishing Editors:

Asia:

China, *Jessie Guo, Assistant Editor* (jessie.guo@springer.com) (Engineering)

India, *Swati Meherishi, Senior Editor* (swati.meherishi@springer.com) (Engineering)

Japan, *Takeyuki Yonezawa, Editorial Director* (takeyuki.yonezawa@springer.com) (Physical Sciences & Engineering)

South Korea, *Smith (Ahram) Chae, Associate Editor* (smith.chae@springer.com) (Physical Sciences & Engineering)

Southeast Asia, *Ramesh Premnath, Editor* (ramesh.premnath@springer.com) (Electrical Engineering)

South Asia, *Aninda Bose, Editor* (aninda.bose@springer.com) (Electrical Engineering)

Europe:

Leontina Di Cecco, Editor (Leontina.dicecco@springer.com)
(Applied Sciences and Engineering; Bio-Inspired Robotics, Medical Robotics, Bioengineering; Computational Methods & Models in Science, Medicine and Technology; Soft Computing; Philosophy of Modern Science and Technologies; Mechanical Engineering; Ocean and Naval Engineering; Water Management & Technology)

(christoph.baumann@springer.com)
(Heat and Mass Transfer, Signal Processing and Telecommunications, and Solid and Fluid Mechanics, and Engineering Materials)

North America:

Michael Luby, Editor (michael.luby@springer.com) (Mechanics; Materials)

More information about this series at http://www.springer.com/series/7818

Seong Oun Hwang · Syh Yuan Tan
Franklin Bien
Editors

Proceedings of the Sixth International Conference on Green and Human Information Technology

ICGHIT 2018

Springer

Editors
Seong Oun Hwang
Department of Computer Information and Communications
Hongik University
Seoul, Soul-t'ukpyolsi
Korea (Republic of)

Franklin Bien
School of Electrical and Computer Engineering
Ulsan National Institute of Science and Technology
Ulsan, Korea (Republic of)

Syh Yuan Tan
Multimedia University
Cyberjaya, Malaysia

ISSN 1876-1100 ISSN 1876-1119 (electronic)
Lecture Notes in Electrical Engineering
ISBN 978-981-13-0310-4 ISBN 978-981-13-0311-1 (eBook)
https://doi.org/10.1007/978-981-13-0311-1

Library of Congress Control Number: 2018944434

Printed on acid-free paper

This Springer imprint is published by the registered company Springer Nature Singapore Pte Ltd.
The registered company address is: 152 Beach Road, #21-01/04 Gateway East, Singapore 189721, Singapore

Preface

ICGHIT 2018, the 6th International Conference on Green and Human Information Technology, was sponsored by the Institute of Electronics and Information Engineers (IEIE), Chiang Mai Rajabhat University, and IEEE Seoul Section. The conference was held from January 31 to February 2, 2018, in Chiang Mai, Thailand. The ICGHIT 2018 Committee consisted of 139 members, whose names are listed on the next page.

The conference received 172 high-quality paper submissions, which were peer-reviewed by experts in the area of the paper's subject. The committee accepted 124 of these submissions and classified them into two categories: The first category includes 50 papers that will be published in these proceedings, and the second category does the remaining 74 papers that will be published in the Proceedings of ICGHIT 2018 by the ICGHIT 2018 Committee. The 40 of 50 papers were finally included in this volume of proceedings.

We would like to acknowledge the great efforts of all the members of ICGHIT 2018 Committee. We also thank Springer who supported the idea of publishing this volume of proceedings. We are grateful to the plenary speakers who kindly accepted our invitation: Jaihie Kim (Yonsei University, Korea), Sansanee Auephanwiriyakul (Chiang Mai University, Thailand), Teerakiat Kerdcharoen (Mahidol University, Thailand), and Trinnawat Suwanprik (Chiang Mai Municipality, Thailand). We would like to express our gratitude to all the authors who submitted their work to ICGHIT 2018.

March 2018

Seong Oun Hwang
Syh-Yuan Tan
Franklin Bien

ICGHIT 2018

General Co-chairs

Kyutae Lee	Kongju National Univ., Korea
Hyunsik Ahn	Tongmyong Univ., Korea

Organizing Chair

Seong Oun Hwang	Hongik Univ., Korea

International Advisory Committee

Seungcheon Kim	Hansung Univ., President of IEIE CIS, Korea
Pathamarat Nakanitanon	Vice President of Rajabhat Univ., Thailand
Beongku An	Hongik Univ., Korea
Surasak Noommeesri	Dean of Rajabhat Univ., Thailand
Worajit Setthapun	Dean of adiCET, Thailand

Local Management Co-chairs

Hoojin Lee	Hansung Univ., Korea
Seri Pansang	Rajabhat Univ., Thailand

International Journal Co-chairs

Chansu Lee	Yeungnam Univ., Korea
Jinghong Kim	Sungkyunkwan Univ., Korea

International Cooperation Co-chairs

Franklin Bien	UNIST, Korea
Chan Yodle	Associate Dean of Rajabhat Univ., Thailand
Pyeoung-Kee Kim	Silla Univ., Korea

Publication Co-chairs

Soyoung Rho	Wolsong Books Co., Korea
Panuwat Suwanakul	Rajabhat Univ., Thailand

Finance Chair

Einjeong Hwang	Myungji Hospital, Korea

Administration Co-chairs

Duckki Lee	Yonam Institute of Technology, Korea
Soohyun Park	Kookmin Univ., Korea

Promotion Co-chairs

Moonsik Kang	Gangneung-Wonju National Univ., Korea
Kotchaphun Boonkong	Rajabhat Univ., Thailand

Industrial Cooperation Co-chairs

Phongsathorn Fongta	Rajabhat Univ., Thailand
Sungchul Yu	LG Hitachi Ltd, Korea

Sejong Kim	SJ Telecom, Korea
Chibong Song	Wavers Co., Korea
Bongsang Suh	All4land, Korea

Workshop Co-chairs

Byung Seo Kim	Hongik Univ., Korea
Sang-ug Kang	Sangmyung Univ., Korea
Natthakarn Amartayakul	DEPA, Thailand
Hye Young Kim	Hongik Univ., Korea

Web Chair

Intae Kim	Hongik Univ., Korea

Technical Program Committee Co-chairs

Yongsoo Choi	Sungkyul Univ., Korea
Young-Hoon Park	Sookmyung Women's Univ., Korea
Seong Oun Hwang	Hongik Univ., Korea
Hoon Jin	Kyonggi Univ., Korea

Technical Program Committee Members

Hyeonsang Eom	Seoul National Univ., Korea
Jungwoo Ryoo	Penn State Univ., USA
Chin Ji Jian	Multimedia Univ., Malaysia
Moon S. Lee	Univ. of Luxembourg, Luxembourg
Taekyoung Kwon	Yonsei Univ., Korea
Rajkumar Kettimuthu	Argonne National Laboratory, USA
Noel Tay Nuo Wi	Tokyo Metropolitan Univ., Japan
Seung-Hun Jin	ETRI, Korea
Tekin Bicer	Argonne National Laboratory, USA
Tom Oh	Rochester Institute of Technology, USA
Wai-Kong Lee	Universiti Tunku Abdul Rahman, Malaysia
Le Sun	Nanjing Univ. of Information Science and Technology, China
Sarhan Musa	Prairie View A&M Univ., USA
Shivani Sud	Intel, USA

Tan Tran Duc	Vietnam National Univ., Vietnam
Sweety Chauhan	Microsoft, USA
Wathiq Mansoor	Univ. of Dubai, UAE
Young B. Choi	Regent Univ., USA
Tan Syh Yuan	Multimedia Univ., Malaysia
Byeungwoo Jeon	SungKyunKwan Univ., Korea
Sansanee Auephanwiriyakul	Chiang Mai Univ., Thailand
Eun Sung Jung	Hongik Univ., Korea
Ooi Shih Yin	Multimedia Univ., Malaysia
Tingjun Xie	Univ. of Virginia, USA
Tudor Palade	Technical Univ. of Cluj-Napoca, Romania
Pang Ying Han	Multimedia Univ., Malaysia
Hyung-Woo Lee	Hanshin Univ., Korea
Wonsoo Kim	Texas Instruments, USA
Claudia Linnhoff-Popien	Institut für Informatikm, Germany
Goh Hock Guan	Universiti Tunku Abdul Rahman, Malaysia
Lim Kian Ming	Multimedia Univ., Malaysia
Pham Viet Hung	Vietnam Maritime Univ., Vietnam
Nalini Iyer	KLE Technological Univ., India
Renny Badra	Universidad Simón Bolívar, Caracas, Venezuela
Geong Sen Poh	MIMOS Berhad, Malaysia
Jianping He	Penn State Univ., USA
Junbeom Hur	Korea Univ., Korea
Tee Connie	Multimedia Univ., Malaysia
Thumrongrat Amornraksa	KMUTT, Thailand
Hoang Van Xiem	Vietnam National Univ., Vietnam
Lee Chin Poo	Multimedia Univ., Malaysia
Vineet Kumar	CG Power and Industrial Solutions, India
Chom Kimpan	Panyapiwat Institute of Management, Thailand
Michele Fiorini	The IET Council Chairman, Italy
Roy Chang	Multimedia Univ., Malaysia
Seon Wook Kim	Korea Univ., Korea
Jayabalan Sudharsan	MRIET, India
Bok-Min Goi	Universiti Tunku Abdul Rahman, Malaysia
Muazzam Khattak	NUST, Pakistan
Junseok Lee	ETRI, Korea
Thai Son Nguyen	Tra Vinh Univ., Vietnam
Anil Dubey	Poornima Institute of Engineering and Technology, India
Renny Badra	Universidad Simon Bolivar, Venezuela
Goh Kah Ong Michael	Multimedia Univ., Malaysia
Cheol-Min Park	NIMS, Korea

Contents

AMQP-Based Subscription Group Message Delivery System Design for Laboratory Safety System

Hyun-seong Lee[1], Jae-gwang Lee[1], Jae-pil Lee[1], Ki-su Yoon[1], Woo-chang Kwon[2], and Jae-kwang Lee[1(✉)]

[1] Hannam University, Daejeon, Republic of Korea
{hslee,jglee,jplee,ksyoon}@netwk.hannam.ac.kr, jklee@hnu.kr

[2] KISTI, Daejeon, Republic of Korea
wckwon@kisti.re.kr

Abstract. Laboratory safety management system is a system that uses the data collected from various sensor devices to detect risk situations and send risk alerts to the required devices. However, since the pre-existing system has different sensor devices connected to the gateway, it requires the standard communication protocol to be introduced. In addition, due to the low process performance of the IoT sensor device and the unstable wireless environment, the pre-existing protocol inevitably involves unnecessary header and complicated processing. Due to these reasons, it is quite insufficient to apply such pre-existing protocol to the safety management environment where various sensor devices are connected. In this thesis, based on the standardized AMQP, the MQTT protocol more lightweight than the pre-existing protocol was used to propose the laboratory safety management system that allows a topic-based group message delivery between sensor device and service. It is determined that using AMQP, a standard protocol, conveniently transplantable into diverse platforms will allow networking among diverse sensor devices, and that using the topic-based message queue will allow prompt and efficient sensor data transfer among various services.

Keywords: Message-queue · IoT · AMQP · MQTT

1 Introduction

Due to the technological advancements, it is now possible to produce advanced sensor devices at low cost, and networking between devices has become more active. Recently, as such sensor devices are being supplied, IoT are being applied and utilized more and more in diverse environments. On the other hand, the laboratory safety management environment consists of diverse IoT sensor devices and a number of laboratories involving such diverse IoT sensor devices. The sensor types include camera sensor, infrared sensor and temperature sensor. Diverse types of sensor data existing in each group are stored or serviced on a real-time based through the IoT platform. The applications serviced to the users through the IoT platform demonstrate

S. O. Hwang et al. (Eds.): ICGHIT 2018, LNEE 502, pp. 1–9, 2019.
https://doi.org/10.1007/978-981-13-0311-1_1

types and functions that are more diversified and complicated. In particular, in the laboratory safety management environment, the required services are real-time data streaming and big data storage/analysis [1].

However, the pre-existing laboratory safety management system uses the method where all data are stored in the database or file management system and the required data are brought to the involved service through DBMS or server system [3, 4]. In the system where various devices are connected, this may cause excessive overhead due to the traffic consisting of data requests made from various services, and such complicated structure may not allow various services to access the required resources. In addition, it would be insufficient since the varying communication protocol and message format among heterogeneous devices would require use of message bridging or system integration [2].

Recently, AMQP (Advanced Message Queueing Protocol) technology, the MOM (Message-Oriented Middleware)-based ISO standard protocol, proposed to resolve such complicated data transfer problem faced by various heterogeneous devices has been receiving attention. In this thesis, based on AMQP, the following systems were proposed: an integrated management system that integrates not only diverse devices using the MQTT (Message Queue Telemetry Transport) protocol, but also various groups that constitute such diverse devices, and a system that efficiently receives messages from diverse services. In Sect. 2, the MQTT protocol and MOM (Message-Oriented Middleware) technology are described. In Sect. 3, the design and detailed structure of the proposed system is discussed. In Sect. 4, the conclusion and future research are discussed.

2 Related Research

2.1 MQTT

MQTT (Message Queue Telemetry Transport) was developed by IBM and EUROTECH, and was distributed as an open standard protocol [5]. Since it uses the lightweight protocol structure of the client/server-based published/subscribed message transfer protocol, it is used for the M2M (Machine to Machine) communication or IoT devices with low transfer rate, and is also used for the message exchange between clients on apps such as online messenger services where a lot of messages are exchanged [6–8].

The MQTT architecture is configured as shown in Fig. 1. It runs over TCP/IP or other network protocols, and consists of a lot of clients based on the MQTT protocol and a broker that manages the connection among such clients and transfers the messages [5]. The clients are divided into publishers who transfer messages to the broker, and subscribers who are registered to receive the messages from the broker. Such structure is quite light and prompt in comparison to the pre-existing protocols such as HTTP and HTML [6]. From the perspective of the broker, the publishers are hierarchically listed and they use the concept of 'Topic' which includes their exclusive name, to transfer messages. The broker classifies the clients through their hierarchized topic.

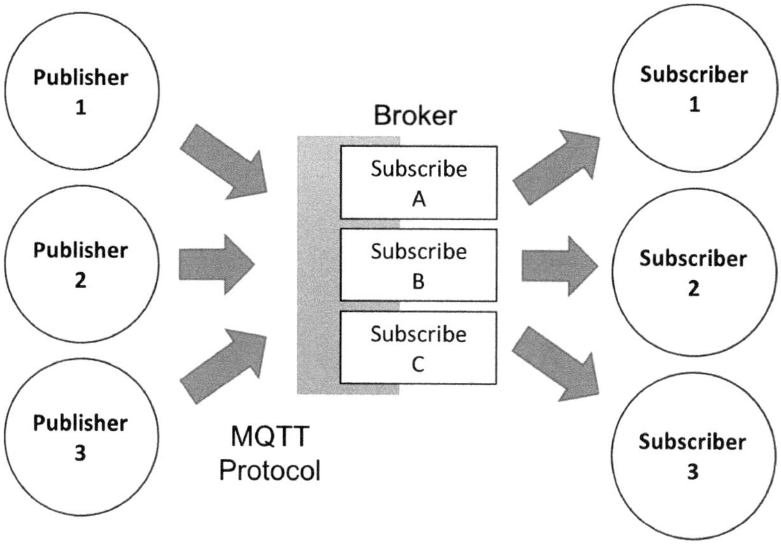

Fig. 1. Architecture of MQTT depends on a MQTT broker

In addition, the subscribers register one or more particular topics. The broker server delivers the messages published with a topic consistent with the subscription to each subscriber.

To guarantee QoS (Quality of Service) in the process of delivering the published/subscribed packet, MQTT provides the following QoS levels for selection.

- QoS Level 0: Any functions provided by MQTT are not added to TCP. It functions to the extent equivalent to TCP's maximum efforts. Whether or not messages reach their destination is not confirmed.
- QoS Level 1: Whether or not messages reach their destination is confirmed. However, there is a possibility that the same message can be repeatedly delivered. Clients may receive the same message more than once.
- QoS Level 2: Messages are accurately delivered to subscribers only once, and the broker stores the delivered messages. There is a possibility that the maximum overhead may occur.

Pre-existing Research in [10, 11], MQTT was compared with the pre-existing web-based communication. As a result, it was found that MQTT showed a transfer rate higher than HTTP, and showed a transfer rate higher even when a large file was transferred. In addition, MQTT provides diverse functions such as keep-alive message and three-level QoS. However, it is difficult to subscribe to a message in case the involved service does not use MQTT [12]. It is necessary to run a separate system to resolve this problem, and doing so may cause unnecessary overhead. In addition, another weakness is that it is difficult to execute additional expansion since clustering is not supported. In case MQTT is used exclusively, it is impossible to execute store-and-forward, and, therefore, it is necessary to have a separate system for storing data in preparation for any network errors [13]. In this thesis, to allow networking among lightweight IoT sensor devices, the MQTT protocol was used and the MOM-based AMQP broker server was separately prepared to compensate for the MQTT-related problems.

2.2 Message Oriented Middleware

MOM (Message oriented middleware) is a system that routes and delivers messages between two or more different clients. Clients call API through the MOM system to send their messages to those who registered to receive the messages. Since the clients who sent their messages no longer are required to get involved in the message transfer process, they are able to work on other tasks. In addition, the received messages can be processed even when a network error occurs. In case MOM is used, the management interface can be added to monitor and expand the performance. Accordingly, clients are able to avoid the message transfer-related problems.

AMQP (Advanced Message Queueing Protocol) was proposed to not only compensate for the weaknesses of the pre-existing MOM system, but also efficiently exchange messages among diverse different systems [14]. AMQP 1.0 was developed by more than 20 leading companies such as Microsoft, Axway and Huawei, and is an international standard approved by ISO and IEC. Its strengths are interoperability, diverse libraries/tools and obsolescence prevention [15]. As the binary protocol to which the latest technology is applied, it features multi-channel, negotiation, asynchronous support, security and efficiency. To be neutral to diverse systems, AMQP must satisfy the following conditions.

- All brokers must operate identically.
- All clients must operate identically.
- The commands transferred on the network must be standardized, and the programming languages must be neutral.

Since it is neutral to diverse systems, it is applicable to the IoT system where various types of devices are connected [16].

The AMQP routing model is configured as shown in Fig. 2. The exchange routes the messages received from the publisher to the queue or another exchange. The message queue is the normally used queue. It temporarily stores messages on the memory or disk and delivers them to the consumers. The queue is directly bound to the exchange through the binding in which the message type is designated. The same queue can be bound to a number of exchanges, and a number of queues can be bound to one exchange.

In [17], based on the ActiveMQ system supporting AMQP, the vocabulary filter classifying the related topics was proposed, and it showed more prompt performance than the pre-existing middleware. In addition, since it supports AMQP, it is possible to integrate heterogeneous devices. However, ActiveMQ can only be realized in one language, and it is difficult to provide the streaming service since it does not feature binding. In [18], the middleware system based on the RabbitMQ system supporting AMQP was compared with the MOM-based Apache Kafka writing queue data on the disk. As a result, Apache Kafka showed higher speed and efficiency but showed lower

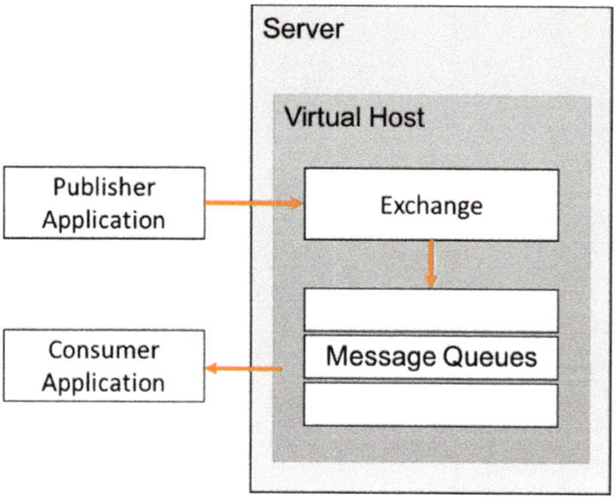

Fig. 2. AMQP routing model process

architecture packaging than AMQP. It was found that it would be more advantageous to use RabbitMQ supporting topic function and AMQP in terms of expandability and integration of diverse devices [18, 19]. In addition, in case a lot of sensor devices are connected, it would be difficult to manage all the traffics received. Accordingly, in the laboratory safety management system environment where diverse sensor devices exist, it would be more reasonable to use the sensor device that uses the MQTT protocol more lightweight than the pre-existing protocol, and to use RabbitMQ supporting AMQP.

3 System Design

The pre-existing laboratory safety management system stores all the sensor data received through the gateway on the database system. The data required by the service can be accessed directly through the database system or connected server. The service requiring streaming relies on the database performance. In addition, since various heterogeneous devices are connected within the laboratory safety management system, it is necessary to construct an interconnecting platform. Figure 3 shows the pre-existing laboratory safety system.

The system proposed in this thesis consists of the followings: the IoT group devices that use AMQP, a standardized protocol, as the basis, the broker that integrates/manages such devices, and the publisher application that provides its services based on the data subscribed to the broker. Figure 4 shows the overall configuration of the system proposed in this thesis.

3.1 MQTT Topic for IoT Devices

MQTT operates as it uses the hierarchically structured topic as its standard. Since all the connected IoT sensor devices can be hierarchically structured, it is possible to

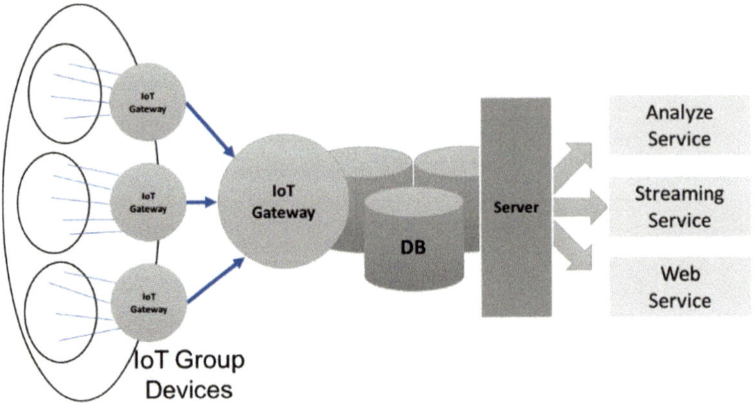

Fig. 3. Pre-existing laboratory safety management system

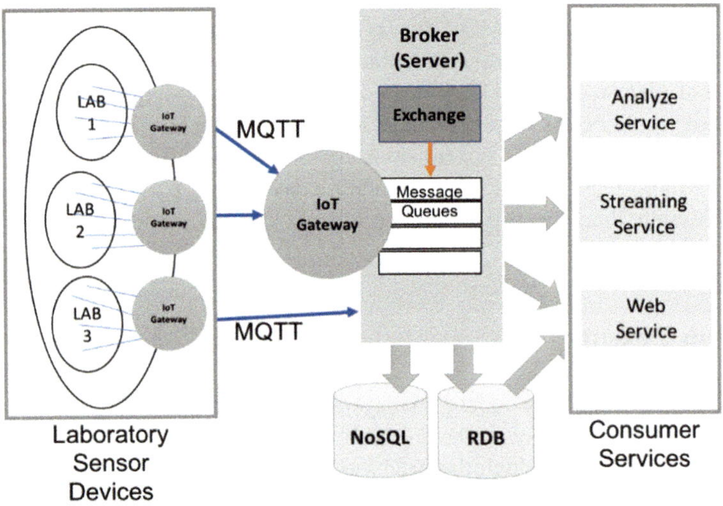

Fig. 4. Sensor devices group messaging system for IoT laboratory

manage them efficiently. Figure 5 show the hierarchical structure of the laboratory sensor devices shown in Fig. 4. Sensor device serving as the publisher uses "/" symbol or "." to express its hierarchy. For example, Fire (sensor node) under LAB1 under Lab Tree shown in Fig. 5 can be expressed as "Lab.Lab1.fire". As far as the MQTT topic structure of the system proposed in this thesis is concerned, just as the tree structure shown in Fig. 5, the laboratories were divided into LAB1, LAB2 AND LAB3 under LAB, and the sensor for the involved laboratory was constructed.

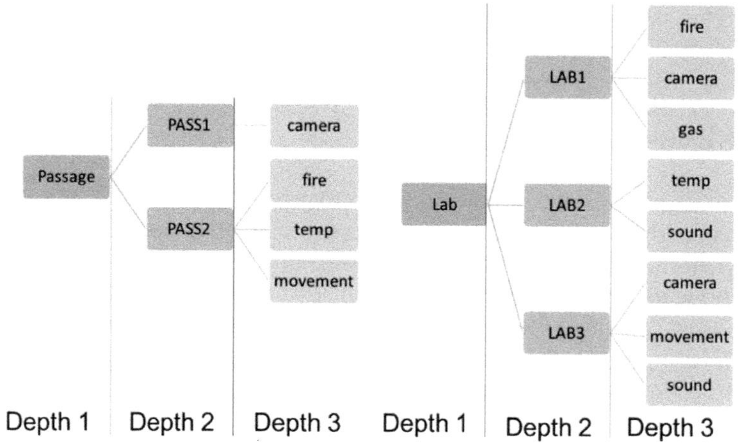

Fig. 5. MQTT topic model design of IoT devices

3.2 AMQP Subscribe for Service

To receive the subscribed messages through the AMQP broker, the topic route must be designated and the subscription must be registered. Table 1 shows scenarios of how AMQP's standard exchange type from the hierarchically structured sensor devices shown in Fig. 5 can be used to diversely subscribe to services. In Table 1, the service represents the applications that are making a request for subscription to the broker, and each depth represents the topic per hierarchy to which the involved application is subscribed.

Table 1. Design of AMQP standard exchange type

Service	Depth 1	Depth 2	Depth 3
Analyze service	Lab	LAB1	Fire, Filter
Streaming service	*	*	Camera, Sound
RDBMS	Lab	*	Fire, Temp, Movement
NoSQL	*	*	*

In Table 1, the subscription to the analyze service is based on the direct exchange method where the name of the topic for binding is used as the queue name. Just as topic://Lab.LAB1.Fire and topic://Lab.LAB1.Filter, the topic can be directly used for binding. The subscription to the streaming service is based on the topic exchange method where the wildcard is used for matching. The streaming service shown in Table 1 uses "*" in the depths 1 and 2 to use various routing keys as reference at once. In the depth 3 it uses the camera and sound topic for matching. The subscription to the RDBMS is based on the fan-out exchange method where all messages are routed to the queue. The subscription to NoSQL is based on the method where regular expressions are used. This method was not used since it involves complicated arithmetic operations that may decrease the speed.

4 Conclusion

As far as the pre-existing laboratory safety management system is concerned, the communication between sensor device and gateway involves complicated processing, and the platform-related problems exist as well. In addition, since it requires use of message bridging or system integration in order to allow messages to be exchanged among various different heterogeneous devices, it is quite insufficient. Moreover, sensor devices operate in the wireless network environment where different processing shows unstable performance. Accordingly, the laboratory safety management system requires the followings: a communication protocol that stably operates in diverse systems, and a platform where heterogeneous devices can be integrated and managed.

In this thesis, based on the standardized AMQP, the IoT device using lightweight MQTT protocol was proposed, and the group messaging system based on the topic among services was proposed as well. The laboratory safety management system proposed in this thesis can be used to efficiently and promptly service the IoT sensor device data in diverse applications. In addition, it is possible to use the standard protocol to easily add new services or systems. In the future research, the plan is to design and implement a system capable of optimizing and parallel-processing AMQP in the proposed system.

Acknowledgment. This research was supported by Basic Science Research Program through the National Research Foundation of Korea (NRF) funded by the Ministry of Science, ICT & Future Planning (NRF-2017R1D1A3B03036130).

References

1. Belli, L., Cirani, S., Ferrari, G., Melegari, L., Picone, M.: A graph-based cloud architecture for big stream real-time applications in the Internet of Things. In: ESOCC 2014, vol. 508 (2015)
2. Qin, Z., Denker, G., Giannelli, C., Bellavista P., Venkatasubramanian, N.: A software defined networking architecture for the Internet-of-Things. In: 2014 IEEE Network Operations and Management Symposium (NOMS), Krakow, pp. 1–9 (2014)
3. Maiti, A.: NETLab: an online laboratory management system. In: IEEE EDUCON 2010 Conference, Madrid, pp. 1351–1358 (2010)
4. Yang, X., Jang, J.M., Jung, H.K.: Advanced Sensor-based control reagent cabinet monitoring system. J. Korea Inst. Inf. Commun. Eng. **21**(1), 199–204 (2017)
5. MQTT Version 3.1.1. http://docs.oasis-open.org/mqtt/mqtt/v3.1.1/os/mqtt-v3.1.1-os.html, last. Accessed 27 Feb 2018
6. Shim, S.H., Kim, H.B.: Internet of Things and MQTT technique. Rev. KIISC **24**(6), 37–47 (2014)

7. Hunkeler, U., Truong, H.L., Stanford-Clark A.: MQTT-S — a publish/subscribe protocol for Wireless Sensor Networks. In: 3rd International Conference on Communication Systems Software and Middleware and Workshops, Bangalore, pp. 791–798 (2008)
8. Kim, S.H., Kim, D.H., Oh, H.S., Jeon, H.S., Park, H.J.: The data collection solution based on MQTT for stable IoT platforms. J. Korea Inst. Inf. Commun. Eng. **20**, 728–738 (2016)
9. Lee, S.H., Kim, H.W., Ju, H.T.: Design of the high-level architecture of mobile integration SNS gateway and the MQTT based push notification protocol. J. KICS **38**, 344–354 (2013)
10. Grgić, K., Špeh I., Heđi, I.: A web-based IoT solution for monitoring data using MQTT protocol. In: 2016 International Conference on Smart Systems and Technologies (SST), Osijek, pp. 249–253 (2016)
11. Chen, H.W., Lin, F.J.: Converging MQTT resources in ETSI standards based M2M platform. In: 2014 IEEE International Conference on Internet of Things (iThings), and IEEE Green Computing and Communications (GreenCom) and IEEE Cyber, Physical and Social Computing (CPSCom), Taipei, pp. 292–295 (2014)
12. Robinson, J.M., Frey, J.G., Stanford-Clark, A.J., Reynolds, A.D., Bedi, B.V.: Sensor networks and grid middleware for laboratory monitoring. In: First International Conference on e-Science and Grid Computing (e-Science 2005), Melbourne, pp. 8–569 (2005)
13. Triawan, M.A., Hindersah, H., Yolanda, D., Hadiatna, F.: Internet of Things using publish and subscribe method cloud-based application to NFT-based hydroponic system. In: 6th International Conference on System Engineering and Technology, Bandung, pp. 98–104 (2016)
14. OASIS Advanced Message Queuing Protocol (AMQP). http://docs.oasis-open.org/amqp/core/v1.0/os/amqp-core-complete-v1.0-os.pdf. Accessed 27 Feb 2018
15. Vinoski, S.: Advanced message queuing protocol. IEEE Internet Comput. **10**(6), 87–89 (2006)
16. Subramoni, H., Marsh, G., Narravula, S., Ping, L., Panda, D.K.: Design and evaluation of benchmarks for financial applications using Advanced Message Queuing Protocol (AMQP) over InfiniBand. In: 2008 Workshop on High Performance Computational Finance, Austin, pp. 1–8 (2008)
17. Lien, Y.C.N., Wu, W.J.: A lexical database filter for efficient semantic publish/subscribe message oriented middleware. In: 2010 Second International Conference on Computer Engineering and Applications, Bali Island, pp. 154–157 (2010)
18. Babovic, Z.B., Protic, J., Milutinovic, V.: Web performance evaluation for Internet of Things applications. IEEE Access **4**, 6974–6992 (2016)
19. Renhak, K., Seitz, J.: User centric multi-purpose messaging framework. In: 2013 International Conference on ICT Convergence (ICTC), Korea, pp. 66–71 (2013)

Attack Probability Analysis on the MTD System

Jang-Geun Ki(✉) and Kee-Young Kwon

Division of Electrical and Electronic and Control Engineering,
Kongju National University, Cheonan 31080, South Korea
{kjg,kky}@kongju.ac.kr

Abstract. Interests in wireless communication technologies have increased in these days, along with rapid prevalence of wireless terminals. Wireless communications have the disadvantage of being vulnerable to security by allowing anyone to receive radio signals by broadcasting wireless signals. There have been lots of researches on MTD (Moving Target Defense) techniques that constantly change the system's vulnerable surfaces over time to protect the wireless communications system from malicious attackers. In this paper, attack analyses have been conducted to improve the security of the wireless communication systems in which the MTD technologies are applied.

Keywords: Moving Target Defense · Attack vulnerability

1 Introduction

Wireless networks have been widely deployed recently because of ease installation, low cost and high bandwidth. However, due to the nature of the wireless signal transmission, wireless communication systems are inherently vulnerable to security breaches, primarily because of easy access to data signal by the unauthorized attackers.

Many studies have been co ducted to improve the wireless security technologies considerably, but attacks such as sniffing, DOS(Denial Of Service), session hijacking, and jamming, etc., also have increased. Therefore, efficient security measures are needed.

In order to prevent the malicious attacks, much attention has been focused on the MTD (Moving Target Defense) [1–4] technology, which is constantly changing the system's vulnerable surfaces that can be attacked in a variety of ways. Communication systems with the MTD mechanism will reduce system vulnerability due to continual changes in the function of the weak points in the system, thereby increasing the cost of attack and decreasing the aggressiveness of attackers, thereby ultimately increasing the self-defense and resilience of the system. These MTD techniques can be applied to all layers ranging from the physical layer of the communication system to the application layer, and the security can be stronger when the MTD technology applies to multiple layers at the same time.

In this paper, attack success probabilities in case of scanning attack and jamming attack have been mathematically analyzed in the MTD-based wireless communication system.

S. O. Hwang et al. (Eds.): ICGHIT 2018, LNEE 502, pp. 10–13, 2019.
https://doi.org/10.1007/978-981-13-0311-1_2

2 MTD System and Attack Modeling

In order to evaluate the attack success probability on the MTD-based communication system [3], assumptions are as follows:

- There are total N channels in the communication system.
- The sender changes the transmission channel at every T time or at the next time slot right after detecting the attack success.
- Attack occurs at every unit time slot and attacker chooses an attacked channel randomly.
- Once the attack succeeds, the success will continue until the transmission channel of the sender changes.

Figure 1 shows an example of the operation of data transmission channel and attack channel when the sender changes the transmission channel at every T time period during scan attack or at the next time slot right after detecting the attack success during jamming attack.

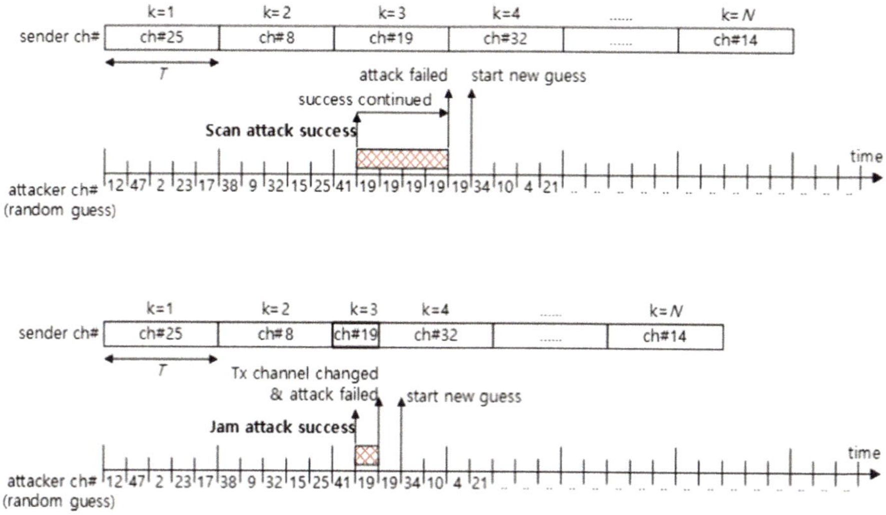

Fig. 1. Example of data channel and attack channel.

3 Attack Success Probability

Depending on the attackers' attack methods, the data sender and receiver may be able to detect whether the attack is successful or not. As an example, scanning or monitoring attack is very difficult to detect while jamming attack can be detected easily. Detecting the attack can affect the sender's decision to change the transmission channel and result in the different attack success probability. Therefore, we consider two kinds of attacks, scanning and jamming attacks. At each time slot, the attack is considered as a success if the data transmission channel of the sender and the selected attack channel of the attacker are the same. The attack success probability is defined as the portion of the attack success slots compared to the total communication slots.

3.1 Scanning Attack Success Probability

Success probability of scanning attack is defined as the ratio of the coincidence between the sender's channel and the attacker's channel when the sender transmits data using the randomly selected channel at every determined interval. Scanning attack success probability is calculated in the following expression.

$$P_{scan} = \frac{\sum_{i=1}^{T} \left(\frac{N-1}{N}\right)^{i-1} \left(\frac{1}{N}\right)(T - i + 1)}{T} \tag{1}$$

3.2 Jamming Attack Success Probability

Success probability of jamming attack is defined as the same way as the scanning attack success probability except that the sender can change its channel right after the attack success as well as at every determined interval. In this case, attacker should newly guess the sender's channel at every time slot because the sender will change its channel right after the attack success.

Jamming attack success probability can be easily obtained as shown below.

$$P_{jam} = \frac{1}{T} \tag{2}$$

3.3 Results

Figure 2 shows the attack success probability of the random scanning attack and the random jamming attack according to the change of the T time.

As shown in the figure, in the case of random scanning attack, the bigger T values result in the increase of attack success probability. On the other hand, in the random jamming attack case, the change of the T value does not affect the attack success probability.

In both attack cases, the higher the N value, the smaller the attack success probability.

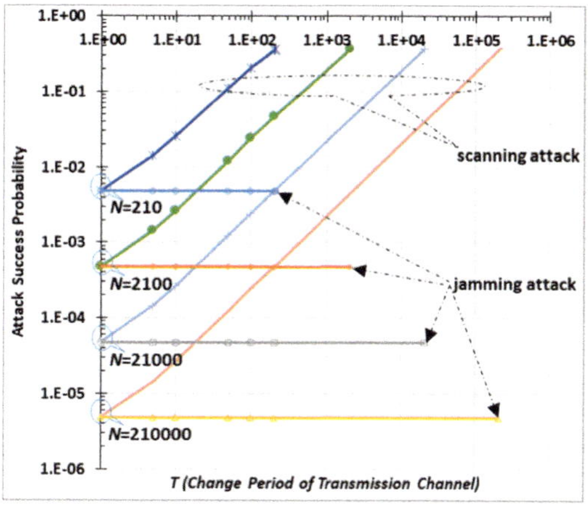

Fig. 2. Random scanning/jamming attack success probability.

4 Conclusion

With the increase of mobile terminals, wireless technologies have been developed rapidly. In order to develop a resilient wireless communication system, a MTD-based approach changes the radio parameters dynamically so that the vulnerability of the system can be protected from the malicious attack.

In this paper, random scanning and jamming attack success probabilities have been analyzed to improve the security of the wireless communication systems in which the MTD technologies are applied.

In the scanning attack, the bigger the T value (period of changing the sender's channel), the higher the attack success probability. On the other hand, in the jamming attack, the change of the T value does not affect the attack success probability. In both attack cases, the higher the N value (total number of channels to be used by the sender), the smaller the attack success probability.

References

1. Casola, V., Benedictis, A.D., Albanese, M.: A moving target defense approach for protecting resource-constrained distributed devices. In: IEEE 14th International Conference on Information Reuse and Integration (IRI), pp. 22–29. IEEE, San Francisco (2013)
2. Kampanakis, P., Perros, H., Beyene, T.: SDN-based solutions for moving target defense network protection. In: IEEE 15th International Symposium on a World of Wireless Mobile and Multimedia Networks (WoWMoM). IEEE, Sydney (2014)
3. Ki, J.G.: Performance analysis of SDR communication system based on MTD technology. J. Inst. Internet Broadcast. Commun. (JIIBC) **17**(2), 51–56 (2017)
4. Jajodia, S., Ghosh, A.K., Subrahmanian, V.S., Swarup, V., Wang, C., Wang, X.S.: Moving Target Defense II: Application of Game Theory and Adversarial Modeling. Springer, New York (2013). ISBN 978-1-4614-5416-8

Extraction Efficiency of Three-Striped LED

Kee-Young Kwon(✉) and Jang-Geun Ki

Division of Electrical and Electronic and Control Engineering,
Kongju National University, Cheonan 31080, South Korea
{kky, kjg}@kongju.ac.kr

Abstract. In a typical GaN based LED structure, the extraction efficiency is about 37.3% when Fresnel reflection is the only mechanism to prevent out-coupling of light. But if the electrode absorption is assumed as 100%, the extraction efficiency is about 26.8%. In a three-striped LED, the extraction efficiency is about 40.7% when Fresnel reflection is the only mechanism to prevent out-coupling of light. But if the electrode absorption is assumed as 100%, the extraction efficiency is about 23.5%.

Keywords: LED · Extraction efficiency

1 Introduction

Light Emitting Diodes (LEDs) are widely used in many commercial products. As a lighting source, LED has some advantages in comparison with incandescent lamps; longer operational lifetime, improved color purity, higher electro-optic efficiency, particularly for colored light, etc.

In the field of general LED application, devices should have the following properties (i) high efficiency and (ii) high power capability. Therefore internal quantum efficiency and light extraction efficiency should be maximized. But owing to the high refractive index of semiconductors, the total internal reflection is occurred on a planar semiconductor-air interface if the angle of incidence is sufficiently large. As a result of total internal reflection, light can be trapped inside the LED die. Light trapped in the die will eventually be reabsorbed by the substrate, active layer, cladding layer, or by a metal electrode.

One of new technologies in high power LEDs is thin-film flip-chip technology [1]. Roughened n-GaN surface improves out-coupling of light. Also photonic crystal in the surface of LED chip enables light extraction through scattering of wave-guided light into radiation modes [2]. And die geometry has a large impact upon extraction efficiency. To date, the most efficient geometry is the truncated inverted pyramid [3].

In this paper, in order to study the effect of tapered three-striped LED structure on the extraction efficiency, the three-striped LED has been compared with a typical LED structure. Also the effect of metal electrode absorption on the extraction efficiency has been compared in these two LEDs.

S. O. Hwang et al. (Eds.): ICGHIT 2018, LNEE 502, pp. 14–18, 2019.
https://doi.org/10.1007/978-981-13-0311-1_3

2 Simulation Model

2.1 Ray Tracing

The external quantum efficiency of an LED, η_{ext}, is the product of the carrier injection efficiency, η_{inj}, and the internal quantum efficiency, η_{int}, and the extraction efficiency, η_{extrac}, that is,

$$\eta_{ext} = \eta_{inj} \times \eta_{int} \times \eta_{extrac}. \tag{1}$$

Extraction efficiency depends on absorption within the substrate material, chip geometry, and relative refractive indices of the die, optical coupling material and transparent package. A unique feature of their ray tracing technique is the geometrical treatment combined with a wave optics approach, taking into account refraction, reflection, absorption and interference effects. Snell's law is applied at the boundary between two dielectric materials with different refractive indices n_i and n_t, as in (2). The angles θ_i and θ_t used in (2) are measured with respect to the surface normal, shown in Fig. 1.

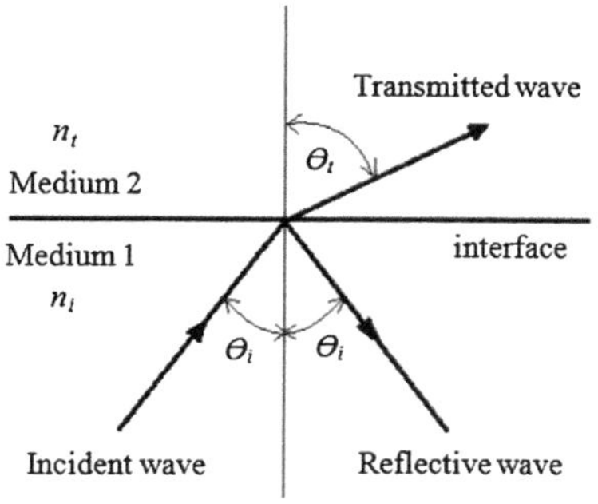

Fig. 1. Reflected and refracted light ray at the boundary between two media with refractive indices n_i and n_t, where $n_i > n_t$.

$$n_i \sin \theta_i = n_t \sin \theta_t. \tag{2}$$

The critical angle θ_c for total internal reflection is obtained using $\theta_t = 90°$, that is,

$$\theta_c = \sin^{-1}(n_t/n_i.) \tag{3}$$

For all angles of incidence $\theta_i > \theta_c$, total internal reflection occurs. Photons are generated in the active layer within the die and propagate in all directions. However, only a small fraction of them is able to pass through the boundary within the small cone. The remainder is reflected from the surface. Outside of the critical cone angle, photons experience total internal reflection and do not contribute to useful light. The light is still subject to Fresnel reflection at the semiconductor-air interface. The Fresnel equations determine the reflection and transmission of light incident on an interface of two media with different indices of refraction. The reflection coefficient for perpendicularly polarized light is given by

$$R- = [n_i \cos\theta_i - n_t \cos\theta_t]/[n_i \cos\theta_i + n_t \cos\theta_t]. \tag{4}$$

The transmission coefficient for perpendicularly polarized light is as follows.

$$T- = 2n_i \cos\theta_i/[n_i \cos\theta_i + n_t \cos\theta_t]. \tag{5}$$

The reflection coefficient for parallel polarized light is given by

$$R_{||} = [n_t \cos\theta_i - n_i \cos\theta_t]/[n_i \cos\theta_t + n_t \cos\theta_i]. \tag{6}$$

The transmission coefficient for parallel polarized light is given by

$$T_{||} = 2n_i \cos\theta_i/[n_i \cos\theta_t + n_t \cos\theta_i]. \tag{7}$$

3 Results

In order to study the effect of tapered stripe structure on the extraction efficiency, the three-striped LED has been compared with a typical LED structure.

Cross section of a typical LED and its geometrical structure parameters used for computer simulation are shown in Fig. 2 and Table 1. Light is generated in the emitting layer of an LED and emitted into all directions. If the incidence angle of the ray is smaller than the critical angle of total internal reflection, then the light is coupled out. The extraction efficiency is about 37.3% when Fresnel reflection at the semiconductor-air interface is the only mechanism to prevent out-coupling of light, that is, there are no metal electrodes. But the p- and n-metal electrodes absorb photons and reduce the light extraction efficiency. If the electrode absorption is assumed as 100%, the calculated extraction efficiency of a typical LED structure is about 26.8%.

Cross section of the three-striped LED and its geometrical structure parameters used for computer simulation are shown in Fig. 3 and Table 2. The extraction efficiency of the three-striped LED is about 40.7% when Fresnel reflection at the semiconductor-air interface is the only mechanism to prevent out-coupling of light, that is, there are no electrodes. But the p- and n-metal electrodes absorb photons and reduce the light extraction efficiency. The effect of p- and n-metal electrode absorption on the light extraction efficiency has been studied. If the electrode absorption is assumed as 100%, the calculated extraction efficiency of the three-striped LED structure is about 23.5%.

If there are no metal electrodes absorbing the photons emitted from the active layer, the tapered striped structure enhances the light extraction. So the extraction efficiency of the three-striped LED is higher than that of a typical LED.

When a photon emission point is located under the p-electrode, the light extraction efficiency is decreased. Therefore when the p- and n-metal electrode absorption is

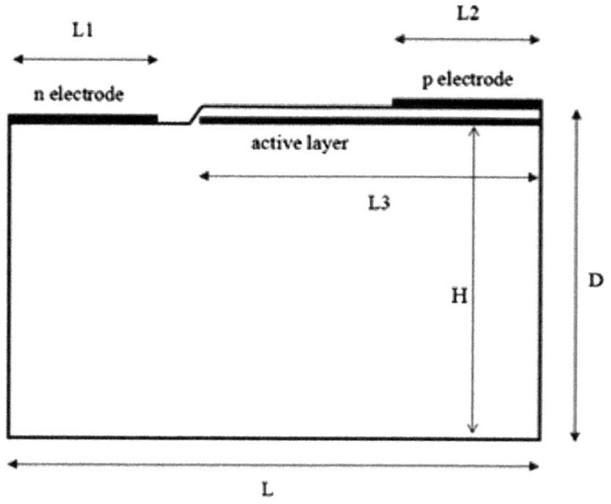

Fig. 2. Structure of a typical LED used for computer simulation with refractive indices n_i and n_o, where $n_i > n_o$.

Table 1. Structure parameters of an LED shown in Fig. 2.

Parameter	Value
n-electrode width, L1 [um]	90
p-electrode width, L2 [um]	90
Active Layer width, L3 [um]	200
Active layer height, H [um]	100
Die width, L [um]	310
Die height, D [um]	103
Inner refractive index, n_i	3.45
Outer refractive index, n_o	1.0

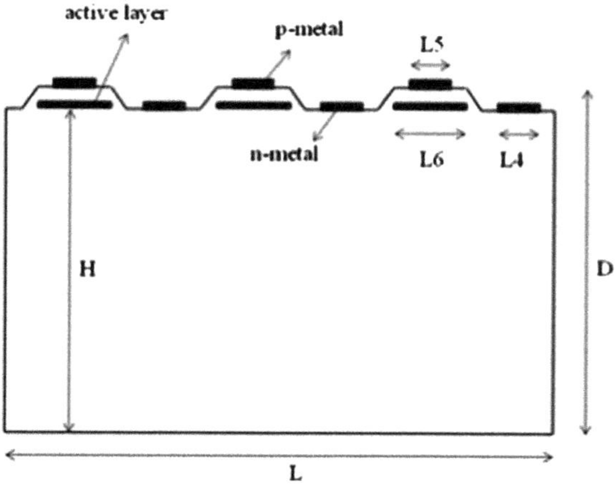

Fig. 3. Structure of the three-striped LED used for computer simulation.

Table 2. Structure parameters of an LED shown in Fig. 3.

Parameter	Value
n-electrode width, L4 [um]	30
p-electrode width, L5 [um]	30
Active layer width, L6 [um]	50
Active layer height, H [um]	100
Die width, L [um]	310
Die height, D [um]	103
Inner refractive index, n_i	3.45
Outer refractive index, n_o	1.0

assumed as 100%, the light extraction efficiency of the three-striped LED is lower than that of a typical LED.

4 Conclusion

In a typical GaN based LED structure, the extraction efficiency is about 37.3% when Fresnel reflection is the only mechanism to prevent out-coupling of light. But if the electrode absorption is assumed as 100%, the extraction efficiency is about 26.8%. In a three-striped LED, the extraction efficiency is about 40.7% when Fresnel reflection is the only mechanism to prevent out-coupling of light. But if the electrode absorption is assumed as 100%, the extraction efficiency is about 23.5%.

References

1. Shchekin, O.B., Epler, J.E., et al.: High performance thin-film flip-chip InGaN–GaN light-emitting diodes. Appl. Phys. Lett. **89**, 071109 (2006)
2. Erchak, A.A., Ripin, D.J., et al.: Enhanced coupling to vertical radiation using a two-dimensional photonic crystal in a semiconductor light-emitting diode. Appl. Phys. Lett. **78**, 563–565 (2001)
3. Krames, M.R., Ochiai-Holcomb, M., et al.: High-power truncated-inverted-pyramid ($Al_x/Ga_{1-x}/)_{0.5}/In_{0.5}$/P/GaP light-emitting diodes exhibiting >50% external quantum efficiency. Appl. Phys. Lett. **75**, 2365–2367 (1999)

Performance Comparison of Internet 4G Service Providers in Bangkok, Thailand

Suttisak Jantavongso(✉) and Sanon Chimmanee

College of Information and Communication Technology, Rangsit University, Pathumthani, Thailand
suttisak.j@rsu.ac.th

Abstract. This paper presents the methodology employed and the findings of the 4G technologies performance for the Internet services in Bangkok, Thailand. The performance of the services was tested both from the stationary and the moving environments. The measurements comprised of the latency, user data rates, and speed tests. The performance comparisons were conducted among three major mobile network operators namely: AIS, DTAC and TrueMoveH during December 2016. In terms of the moving environment, TrueMoveH provided the best result for the latency tests (foreign website) and AIS was the best for the domestic website. Whereas, TrueMoveH and AIS were offered the similar user data rate results in downloading, AIS was offered the better results in uploading. Moreover, the speed test performance revealed that TrueMoveH was offered the best downloading speed. Both TrueMoveH and AIS offered the similar performance when uploading. In relation to the stationary environment, AIS offered the best result for the latency tests for the domestic website. DTAC and TrueMoveH offered the similar result for the foreign website. TrueMoveH offered the best performance in downloading, while AIS performed the best in uploading for the user data rates. Similarly, TrueMoveH gave the best performance in the speed tests in downloading, while AIS performed the best in uploading. The findings from this study assist the operators to benchmark their current Internet services. At the same time, it can also assist the consumers to select the appropriate services to meet their needs without a commercial bias.

Keywords: Mobile network operator · Latency · User data rate
Speed test

1 Introduction

The Fourth Generation (4G) technologies have capabilities to offer Internet services with broadband capacity which able to support multimedia services. Thus, the 4G technologies are envisioned as the solution for the Thailand's lagged behind other countries in the region. Whereas Thailand was one of the last few countries in Southeast Asia to introduce 3G capabilities [1], the first 1800 MHz spectrum auction for 4G service was completed on 12th November 2015. Following this, the second 900 MHz auction was completed on 15th December 2015 [2]. The three major mobile network operators offering 4G services in Thailand are AIS, DTAC and TrueMoveH [3].

S. O. Hwang et al. (Eds.): ICGHIT 2018, LNEE 502, pp. 19–24, 2019.
https://doi.org/10.1007/978-981-13-0311-1_4

4G technologies offer Internet services with rapid data access to support the growth of mobile devices to Thai consumers. The Internet services provided by 4G networks are the dominant technologies. The usage of the Internet in Thailand is expected to continue to increase. Thai Internet users are currently using their mobile devices instead of their desktop devices due to the mobility and ready connectivity while on the go reasons. Network expansion to the 4G network has sped up data transmission rates of the mobile phone networks and provided more convenient and faster access to video and audio files. The download speed of the 4G networks in Thailand are approximately 100 Mbps and upload speed of 50 Mbps [3].

People in Bangkok spend approximately 34 h per week on the Internet. In additional to accessing emails and searching for information, social media is gaining more popularity as an activity of choice. The provision of 4G services by mobile operators has enabled users to always use Internet services while travelling, shopping, eating out in restaurants, or going on excursions. The Bangkok Mass Transit System (BTS) is an elevated rapid transit system in Bangkok. BTS is also known as the skytrain. There are approximately 650,000 commuters travelling by the BTS skytrain each day with an increase of 10% per year. These commuters spend most of their time on their mobile devices using Internet services while travelling [4].

The main objective is to evaluate the performance comparison of 4G mobile network operators in Thailand within Bangkok from the stationary and moving environments. The paper provides information to Thai users in deciding whether they are satisfied with the provided services or not. This would benefit 4G users in Thailand in selecting their mobile network providers without a commercial bias intention.

2 Literature Review

4G is a key growth driver for mobile network operators in Thailand [2]. Since the 4G spectrum auctions in 2015, the network operators have invested heavily in installing their network base stations and expanding their network coverage. For example, AIS has spent 14.5 billion baht on installing 7,000 base stations. TrueMoveH spent 56 billion baht to launch its 4.5G network expecting to cover 97% of the Thai population. AIS has 20 of 38 million subscribers surfed Internet via mobile devices. TrueMoveH networks offer dual-band coverage on the 900 MHz and 1800 MHz band. TrueMoveH installs 13,500 base stations on the 900 MHz network; and 3,000 to 4,000 base stations on the 1800 MHz network providing services in Bangkok and major provinces. TrueMoveH occupies a total of 55 MHz of bandwidth which comprising of 15 MHz on the 2100 MHz band, 15 MHz on the 850 MHz band, 15 MHz of 1800 MHz and 10 MHz of the 900 MHz band [5].

Mobile phone penetration in Thailand is expected to expand 3 to 4 times by 2020 [5]. The emerging technologies such as Internet of Things (IoT) will be deployed in education, telemedicine, logistics and security sectors. Thai mobile network operators will require more bandwidth to serve with the growing bandwidth-intensive services, demanded by the proliferation of mobile devices and faster speed of wireless broadband networks. Having a variety of mobile spectrum bands will enable the mobile network operators to greatly manage frequencies and data traffic flows covering all demands. The technology transitions and the 4G spectrum auctions have increased Thai industry innovation and competitiveness. It provides benefits to Thailand's economy. The enhanced bandwidth and improved data rate on a 4G network are expected to increase more mobile network subscriber experience for higher data traffic, more data intensive applications, and rich communication services. The transition from 3G to 4G and technology utilization will require large scale network investment. This in turn supports the government's vision for a digital economy [2, 5, 6].

There are various methods to measure the Quality of Service (QoS) of Internet services which operators provided. Examples of these methods are the latency, user data rates, and speed test measurements [1, 3, 4, 7–11]. Haryadi and Andina [11] studied the File Transfer Protocol (FTP) services for their QoS measurement. Almutairi et al. [9] evaluated the QoS with Round Trip Time (RTT), FTP, and WWW. Ahmed [8] used the latency (Ping) and the speed tests in his performance test of 4G networks in Saudi Arabia. Dong et al. [7] tested the QoS on the high-speed environment on high-speed trains in China with the RTTs. Hence, RTT measures the time it takes for a data packet to travel from a source across a network to a destination and back. The FTP was used to download a single file. Thus, WWW browsing was a popular application of the Internet in Kuwait.

For Thailand, the 4G mobile networks hold a promise of performance that is comparable or even better than the broadband services provided by 3G or landline Wi-Fi access. However, even though the technical limitations are high, due to the complexities associated with the mobile data networks and multimedia services such as Voice over Internet Protocol (VoIP) on mobile devices, video streaming, content downloading, and diversified services and applications [6]. Moreover, there are online game and social media applications on the mobile platform. Each of the applications requires different quality and means of service provision. For example, multiplayer games require rapidity and real-time response, while social media applications like Facebook and Instagram only require adequate bandwidth for video or photo uploading [6]. Fundamentally, 4G technology is designed to provide real-time, delay-sensitive multimedia to support a different type of experience. Therefore, users in Thailand are also demanded to experience the highest possible level of quality of 4G services. This study, the latency tests, user data rates, and speed tests were the QoS measurements were used to compare the performance of three mobile operators.

3 Measurement Methodology

Five measurement sites in Bangkok were selected for the stationary environment. These locations were selected as they are the most congested and highest demand sites of the 4G usages in Bangkok. These five testing sites were adequate in the performance testing. There were existing studies which had conducted their tests in these five locations [1, 10]. The distance between the test points and the base stations of the three network operators was almost similar in each of the test sites. The selected measurement sites were: Siam Paragon, Central Plaza Lardprao, Pantip Plaza, Queen Sirikit National Convention Center. and AIS Tower. In terms of the moving environment, the Sukhumvit Line travelling between Mo Chit station and On Nut station.

The tests were conducted during December 2016 on weekdays and weekends. The tests were conducted during mornings, afternoons and evenings. The morning period was from 8AM–11.59PM. The afternoon period was from 12PM–15.59PM, and 6PM–8PM in the evening period. The tests were performed at one location each day. The findings were obtained using the means from the three measurements in each period.

Three measurement tests covering the Internet services provided by the three mobile network operators were measured. These tests included the latency, user data rates, and speed test measurements [1, 3, 4, 10].

The latency is measured as the RTT. By utilizing the Internet Control Message Protocol (ICMP) ping command, it is possible to time the latency. In total, there were two tested websites in conducting the latency measurements. The domestic website was "www.kapook.com" and the foreign website was "www.google.com".

All data rate tests were performed during off peak and peak hours in order to monitor the realistic usages of the mobile networks. In all locations, downloading (DL) and uploading (UL) file transfers are performed. The file transfers are performed via the FTP, and the average FTP DL or UL throughput is computed per file transfer. FTP test with AndFPT application testing was performed to transfer 5 MB file by downloading, uploading, and collecting the data. Tests are in a unit of seconds.

The speed test measures the network downloading and uploading speed. The test performs three measurements, download speed, upload speed and RTT resulting in mbps and ms respectively via the "Speedtest by Ookla" application.

4 Test Results and Discussion of the Results

The findings of the latency test results indicated that for the stationary locations, AIS was leading other operators on the average RTT values when tested with the domestic website. For the foreign website, TrueMoveH and DTAC had achieved the better performance overall. In terms of the moving locations, TrueMoveH provided the best result for the latency tests (the foreign website) and AIS was the best for the domestic website.

For the stationary locations, TrueMoveH offered the best performance in downloading, while AIS performed the best in uploading for the user data rate tests. TrueMoveH and AIS were offered the similar user data rate results in downloading, AIS was offered the better results in uploading under the moving locations.

For the outcome of the speed tests for the stationary locations, TrueMoveH gave the best performance in the speed tests in downloading, while AIS performed the best in uploading. In terms of the moving locations, the speed tests revealed that TrueMoveH was offered the best downloading speed. Both TrueMoveH and AIS offered the similar performance when uploading.

The findings indicated the better performance in terms of the latency, user data rates, and speed test measurements from 4G on the stationary locations, except the downloading in the user data rates. The best performances in each test were highlighted and presented in each of the following Tables. In line with the studies by Sauter [12] and Ahmed [8], these mobile network operators were able to provide the better services on the stationary locations, compare to the moving environments (Table 1).

Table 1. Average values of 4G measurements.

State (4G)	Latency (ms)		User data rate (ms)		Speed test (mbps)	
	domestic	foreign	upload	download	upload	download
Stationary	49.72	69.72	12.86	68.82	27.93	31.35
Moving	52.60	84.42	13.60	22.87	18.80	18.87

In terms of the stationary locations, TrueMoveH and AIS offered the similar performance. TrueMoveH has the edge over AIS in downloading and the connection to the foreign website, while AIS provided the better performance for the uploading and connecting to the domestic website. Similar results can be concluded for the moving environments (see Table 2).

Table 2. Mobile network operators' comparison.

	Stationary		Moving	
	Domestic	Foreign	Domestic	Foreign
Latency	AIS	DTAC/TrueMoveH	AIS	TrueMoveH
	Download	**Upload**	**Download**	**Upload**
User data rates	TrueMoveH	AIS	TrueMoveH/AIS	AIS
Speed test	TrueMoveH	AIS	TrueMoveH	TrueMoveH/AIS

Our findings were resulted in accordance with the studies by Malisuwan et al. [2, 5, 6, 13] that since the 4G spectrum auctions in November and December 2015, both AIS and TrueMoveH have invested heavily in installing their network base stations and expanding their network coverage. The performance of their 4G mobile networks are in line with the investment made.

5 Conclusion

For Thailand, 4G mobile network is a high-speed wireless network technology capable of a much higher data transmission rate, with signal quality superior to that of the existing 3G network. The findings indicated the mobile network operators provided the better performance in terms of the latency, user data rates, and speed test measurements for their 4G services for the stationary locations. However, there was no clear winner of the performance of these three operators. It is all depended on the intended usage. Whereas, 4G is implementing in Bangkok and the rest of Thailand, it would be an interesting area for a future work to examine the performance test for pending 5G in the network congestion areas in Bangkok including the BTS stations.

Finally, the authors express the sincerest thanks to Aiyaret Boonyaruang and Prapasri Kiti-chalermkiet for their contributions to this research.

References

1. Chimmanee, S., Jantavongso, S.: The performance comparison of third generation (3G) technologies for internet services in Bangkok. J. Inf. Commun. Technol. **15**, 1–31 (2016)
2. Malisuwan, S., Kaewphanuekrungsi, W., Tiamnara, N., Milindavanij, D., Kaewphanuekrungsi, W., et al.: Licensing of spectrum for mobile service in the frequency band of 900 MHz in Thailand. Int. J. Innov. Res. Sci. Technol. (IJIRST) **2**, 39–45 (2016)
3. Kitichalermkiet, P., Chimmanee, S.: The performance comparison of fourth generation (4G) technologies for internet services in the most congestion areas in Bangkok. In: The 4th SAU National Interdisciplinary Conference 2017 (SAUNIC 2017), Nonthaburi, pp. 1478–1485 (2017)
4. Boonyaruang, A., Chimmanee, S.: The performance comparison of 4G mobile phone on the Bangkok Mass Transit System (BTS) in Bangkok. In: The 4th SAU National Interdisciplinary Conference 2017 (SAUNIC 2017), Nonthaburi, pp. 1495–1501 (2017)
5. Malisuwan, S., Kaewphanuekrungsi, W.: Analysis of roadmaps and trends for mobile communication technology in Thailand. Int. J. Adv. Res. Eng. Technol. (IJARET) **7**, 68–79 (2016)
6. Malisuwan, S., Milindavanij, D., Kaewphanuekrungsi, W.: Quality of service (QoS) and quality of experience (QoE) of the 4G LTE perspective. Int. J. Future Comput. Commun. **5**, 158–162 (2016)
7. Dong, P., Du, X., Zheng, T., Zhang H.: Improving QoS on high-speed vehicle by multipath transmission based on practical experiment. In: The 2015 IEEE Vehicular Networking Conference (VNC), pp. 32–35 (2015)
8. Ahmed, M.H.: Performance test of 4G (LTE) networks in Saudi Arabia. In: The 2013 International Conference on Technological Advances in Electrical, Electronics and Computer Engineering (TAEECE), Konya, pp. 28–33 (2013)
9. Almutairi, A.F., Al-Jame, F., Abdullah, A.: GPRS/EDGE/3G/HSDPA networks performance measurements in the State of Kuwait. In: The 5th IEEE GCC Conference & Exhibition, Kuwait City, pp. 1–3 (2009)
10. Chimmanee, S., Jantavongso, S., Kantala, S.: The mobile technologies performance comparison for Internet services in Bangkok. In: the 7th International Conference on Information Technology and Electrical Engineering (ICITEE), Chaing Mai, pp. 337–342 (2015)
11. Haryadi, S., Andina, R.: QoS measurement of file transfer protocol services in 3G networks using aggregation method. In: The 7th International Conference on Telecommunication Systems, Services, and Applications (TSSA), Bali, pp. 107–110 (2012)
12. Sauter, M.: 3G, 4G and Beyond: Bringing Networks, Devices and the Web Together. Wiley, Chichester (2013)
13. Malisuwan, S., Teppayayon, O., Madan, N., Sivaraks, J., Kaewphanuekrungsi, W.: Analysis of mobile spectrum management in Thailand to move towards a knowledge based economy: a regulatory review. J. Adv. Manag. Sci. **1**, 288–298 (2013)

Performance Evaluation of LoRaWAN on Campus Environment

Dong Hoon Kim, Jae-Seong Jo, and Eun-Kyu Lee(✉)

Incheon National University, Incheon, Republic of Korea
eklee@inu.ac.kr, picopoo@naver.com

Abstract. LoRaWAN is emerging as one of promising wireless technologies for long range IoT communications. At its early stage, its technical specification promises long, reliable communications in ranges up to tens of km (in a rural area). However, its performance of connectivity has not been proved in a real world environment yet. To answer a question of *Does it work well in the real setup?,* this paper evaluates connectivity performance of LoRaWAN communication technology on a real world campus environment. To this end, we have built a testbed system on our campus and ran experiments measuring RSSI, SNR, and PDR. By analyzing experimental results, we evaluate data delivery performance of LoRaWAN technology.

Keywords: IoT · Low power wide area network · LoRa · LoRaWAN
Wireless communication · Testbed

1 Introduction

In the Internet of Things era, it is expected that a number sensor and actuator nodes are deployed and connected. The nodes communicate with the Internet via various wired and wireless communication technologies such as Ethernet, WiFi, Bluetooth, and ZigBee. When considering the volume of deployed nodes, it is impossible to connect them via wire. The WiFi is now widely used for mobile devices such as laptops and smartphone, but its energy usage plan does not fit to energy-constrained IoT devices. In this sense, Bluetooth and ZigBee can be good candidates for IoT communications. However, they can communicate with devices in the vicinity with short radio ranges. In the new era, we expect that devices are everywhere: on the roof, on the street, and in cars. It is challenging and costly to install gateways for Bluetooth and ZigBee every block. To overcome the problem, Low Power Wide Area Network (LPWAN) technology emerges, enabling long range communications with low energy consumption. At its early stage, however, it is not yet proved to operate well in a real world environment [7]. In this paper, we run experiments to evaluate the connectivity performance of an LPWAN technology, LoRaWAN, on a campus environment.

S. O. Hwang et al. (Eds.): ICGHIT 2018, LNEE 502, pp. 25–30, 2019.
https://doi.org/10.1007/978-981-13-0311-1_5

2 Low Power Wide-Area Network

LPWAN is a type of wireless communication network designed to allow long range communications at a low bit rate among things, such as sensors operated on a battery. There are many technologies competing in the LPWAN space, such as LoRaWAN, SigFox, and LTE-M.

2.1 LoRa and LoRaWAN

LoRa (Long Range) is the physical layer or the wireless modulation utilized to create the long range communication link [1]. Many legacy wireless systems use Frequency Shifting Keying (FSK) modulation as the physical layer because of efficient modulation for achieving low power. LoRa is based on chirp spread spectrum (CSS) modulation. The advantage of LoRa is in the technology's long range capability. A single gateway or base station can cover entire cities or hundreds of square kilometers. Range highly depends on the environment or obstructions in a given location, but LoRa has a link budget greater than any other standardized communication technology. So with a minimal amount of infrastructure, entire countries can easily be covered. Another advantage is LoRa uses non-licensed band frequency; 433 MHz, 868 MHz, or 915 MHz.

LoRaWAN defines the communication protocol and system architecture for the network while the LoRa physical layer enables the long-range communication link [2]. The protocol and network architecture have the most influence in determining the battery lifetime of a node, the network capacity, the quality of service, the security, and the variety of applications served by network. Figure 1 illustrates the LoRaWAN protocol stack including three different classes.

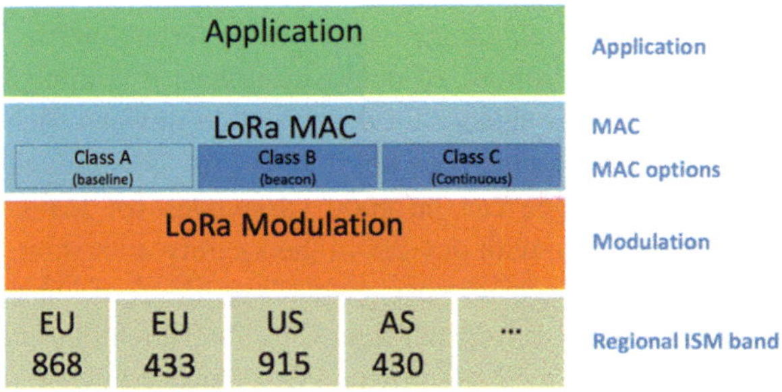

Fig. 1. LoRaWAN protocol [2].

2.2 SigFox and LTE-M

SigFox is a narrowband technology. It uses a standard radio transmission method called BPSK (binary phase-shift keying), and it takes very narrow chunks of spectrum and changes the phase of the carrier radio wave to encode the data. SigFox wireless systems send very small amounts of data (12 bytes) very slowly using standard radio transmission methods. SigFox uses the non-licensed band frequency like LoRa.

LTE-M is LTE-MCT (Machine Type Communication) standardized by 3GPP (3rd Generation Partnership Project). LTE-M uses Cat.1 terminal defined in Released 8 and adds the Power Saving Mode (PSM) function defined in Release 12. This technology uses licensed band frequency and does not require a new network installation. Table 1 compares three LPWAN technologies in a list of criteria.

Table 1. Comparison of LPWAN technologies

	LoRa	SigFox	LTEM
Standard	LoRa Alliance, IEEE 802.14.5 g	ETSI	3 GPP
Frequency band	ISM	ISM	Licensed
Bandwidth	125/250/500 kHz	200 kHz	1.4 MHz
Modulation	CSS	UL: D-BPSK, DL: GFSK	QPSK/16 QAM
Data rate (max)	980 bsp–21.9 Kbps (US)	UL: 160 bps, DL: 600 bps	1 Mbps
Battery life	~10 years	~10 years	~10 years
Coverage	~15 km (rural), ~5 km (urban)	~30 km (rural), ~10 km (urban)	~11 km

3 Building a Testbed System

For experiments, we have built a testbed system that consists of two end nodes, a gateway (or access point), and a data server as shown in Fig. 2.

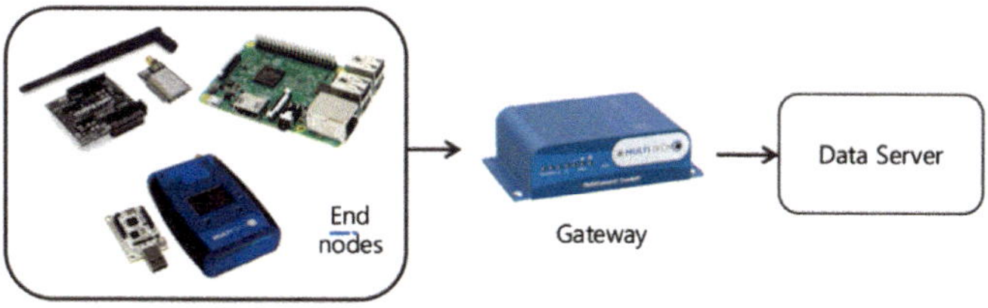

Fig. 2. We have built and deployed a testbed system (end nodes, gateway, and a server)

We build two end nodes. The one is with Arduino Uno R3 boards, a set of sensors, Libelium LoRaWAN 900 module [4], and 4.5 dBi antenna. The LoRaWAN module uses LoRaWAN v1.0, class A. It uses 900/915 MHz ISM frequency band, and TX Power is up to +14 dBm. The module's range is up to 15 km at suburban and 5 km at urban area. The other node is a blue mDot Box that is primarily designed for performance evaluation [5]. It is equipped with four sensors and uses Semtech SX 1272 chip for LoRaWAN and 4.5 dBi Antenna. The box operates in three different modes. In a survey single mode, a user can adjust three parameters of spreading factor, bandwidth, transmission power and then make communications with a gateway. A survey sweep mode is given a list of combination pairs <spreading factor, transmission power>. When communicating with the gateway, the box sweeps the list for performance comparison. In a LoRa demo mode, a user is able to set up a periodic transmission or push a button to send data to the gateway.

The testbed system includes a gateway that is implemented using Multitech's MultiConnect Conduit [6]. It uses Semtech SX 1301 chip, MTAC-LoRa card, and 6 dBi antenna. The MTAC-LoRa card uses an RF spread spectrum technology of Semtech to support a range of 15 km (in rural) and 2 km (in urban). The gateway is also equipped with an Ethernet port with which data is forwarded to the data server. The receiving sensitivity is computed as below, and Table 2 describes examples of sensitivity with popular configurations. Note that DR, SF, BW stand for data rate, spreading factor, and bandwidth, respectively. The gateway is protected by a plastic enclosure, and Fig. 3 shows the gateway built. Upon receiving signal from a node, the gateway collects log of the wireless signal such as receive power as well as sensor records. Then, it forwards the collection to the data server.

Table 2. Sensitivity with popular configuration sets.

DR	SF	BW	Physical DR [bps]	Sensitivity [dBm]
0	10	125	977	−134.0
1	9	125	1758	−131.5
2	8	125	3125	−129.0
3	7	125	5469	−126.5
4	8	500	12500	−123.0

4 Experiments and Preliminary Results

The goal of the first experiment is to test how far the LoRa signal can reach in a real world environment – that is, a test of communication range. To this end, we have deployed the testbed system on a park next to our campus. Since the park is next to sea and have no buildings, but some low trees, it is the best place to test propagation properties of wireless signals. The total distance between a gateway and a node is 600 m, and they are placed at the 1 m height. The node, at every 50 m distance point, sends 100 packets to the gateway that measures and records three values: RSSI (Received Signal Strength Indicator), SNR (Signal to Noise Ratio), and PDR (Packet Delivery Ratio).

4.1 RSSI and Distance

Figure 4 (left) draws the change of the RSSI values with varying distance (solid line). The curve tells that RSSI decreases with the fourth power of the distance in reverse

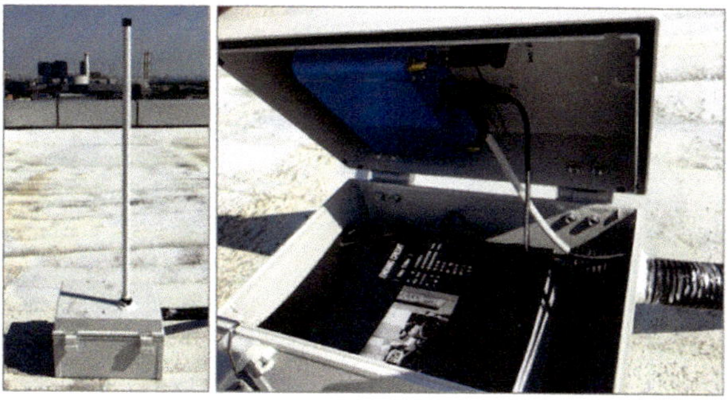

Fig. 3. A gateway is built with LoRa module, antenna, enclousure

proportion. When applying the measured values to the sensitivity equation in Sect. 3, we can compute the average sensitivity of the gateway is −128.8 dBm. From this value and the curve in the previous graph, we can compute that the maximum distance for successful communications becomes 1,600 m. An interesting observation is that the variation of the RSSI values is large enough as shown in Fig. 4 (right). More specifically, the deviation goes up to 5 dB, indicating that RSSI fluctuates in the range of 2σ. This implies that a margin of 10 dB is needed and the range for stable communication decreases down to 900 m.

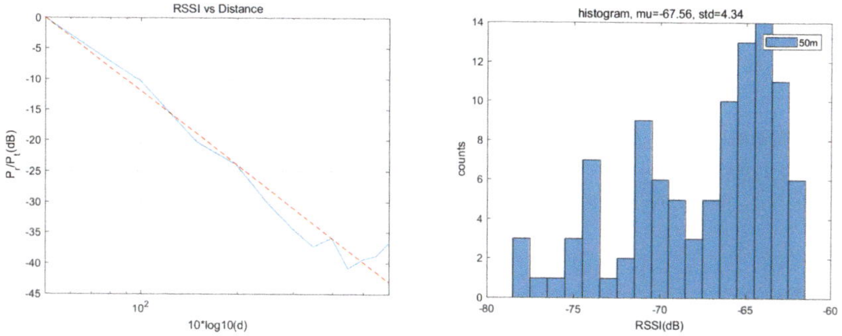

Fig. 4. The RSSI decreases with increasing distance (left). RSSI values at 50 m point are countered and drawn (right).

4.2 Signal to Noise Ratio

Figure 5 (left) shows the change of SNR values with increasing distances. It is not easy to identify any clear relationship between to parameters. But, it is noted that the variation generally increases as the distance increases. To investigate the SNR further, we draw a graph of SNR-PDR as shown in Fig. 5 (right). X, Y axis represent SNR [dB] and PDR [%]. On the right side (greater than 0 dB of SNR), we can observe that PDR

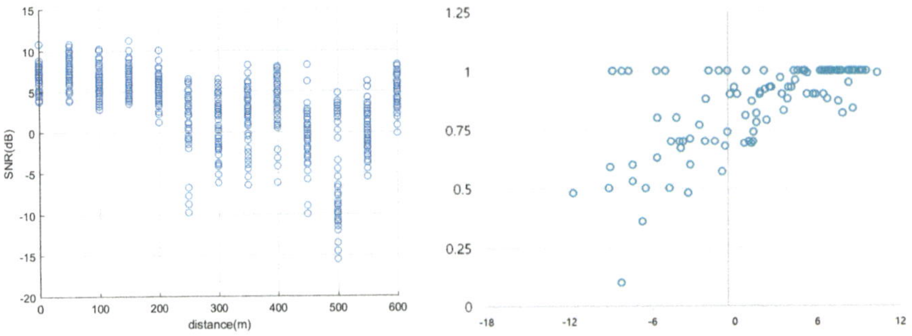

Fig. 5. Distance-SNR points (left). This graph shows the PDR along with varying SNR (right)

is more than 0.75 in general, indicating stable communications. On the other side, PDR fluctuates significantly.

5 Conclusion

To answer a question of *Does LoRaWAN work well in a real world environment?*, this paper has built a testbed system supporting Line of Sight (LoS) on our campus. We also ran experiments to evaluate its connectivity performance for IoT applications. RSSI decreases with the fourth power of the distance in reverse proportion. However, it fluctuates in a significant manner. We also identified the relationship between SNR and PDR. A research question for Our future work is *How to design a LoRaWAN network to cover our campus?*

Acknowledgement. We thank Kyungwoo Kang and Jaehyun Park for their help to make our experiments possible. This research was supported by the Incheon National University Research Grant in 2017. Eun-Kyu Lee and Jae-Sung Jo are co-corresponding authors.

References

1. A technical overview of LoRa and LoRaWAN, LoRa Alliance. https://www.lora-alliance.org/lorawan-white-papers. Accessed 10 Dec 2017
2. Sornin, N., Luis, M., Eirich, T., Kramp, T., Hersent, O.: LoRaWAN specification v1.0.2 (2016)
3. SigFox. https://www.sigfox.com/en. Accessed 10 Dec 2017
4. Libelium LoRaWAN 900. https://www.cooking-hacks.com/lorawan-radio-shield-for-arsduino-900-915-mhz. Accessed 10 Dec 2017
5. MTDOT-BOX-G-915-B, Multi-Tech Systems Inc. https://www.multitech.com/models/99999211LF. Accessed 10 Dec 2017
6. MultiConnect Conduit, Multitech. https://www.multitech.com/models/94557202LF. Accessed 10 Dec 2017
7. Adelantado, F., Vilajosana, X., Tuset-Peiro, P., Martinez, B., Melià-Seguí, J., Watteyne, T.: Understanding the limits of LoRaWAN. IEEE Commun. Mag. **55**(9), 35–40 (2017)

Performance of Nano-crystalline Inductive Couplers Applied to Ship Powerline Communication

Kyung-Rak Sohn(✉), Seung-Ho Yang, and Jae-Hwan Jeong

Korea Maritime and Ocean University, Yeongdo-Gu, Busan 49112, South Korea
krsohn@kmou.ac.kr

Abstract. We have proposed a nano-crystalline inductive coupler that can arbitrarily add or move communication facilities in the ship without physical power line contact. Using a nano-crystalline inductive powerline communication system, we have achieved data rates of more than 12 Mbps between two separate cabins in the training-ship and succeeded in transmitting images on the same powerline condition.

Keywords: Nano-crystalline · Ferrite · Powerline communications Inductive coupler · Capacitive coupler

1 Introduction

The development of power line communication technology has been concentrated on indoor applications of residential buildings, but research on industrial applications is getting less attention. However, inductive powerline communication (PLC) can be used as communication means for various industrial applications [1]. There have been studies to apply inductive PLC to ships [2] and trains [3] using ferrite cores.

Two types of coupling circuits are common, either capacitive or inductive. It should provide the necessary galvanic isolation of the PLC system from the power cable. The capacitive coupling shows the required high-pass filtering characteristics. The signal power is coupled through the displacement current by the capacitor. Conventional narrow-band PLCs are two-wire systems that inject signals between wires using capacitive couplers. It's easy and compact, but requires physical contact with the power cable. In the case of inductive coupling, the signal power is coupled to the power line in a noncontact manner by electromagnetic induction. The main advantage is that the coupling circuit provides isolation from the power source voltage to the signal source without electrical contact and is simple to install and remove [4].

Two representative core materials for inductive coupling unit (ICU) are ferrites and nano-crystalline magnetic materials. Ferrites are ceramic-like materials with a typical saturation flux density about 0.38 T for high permeability materials of about 10,000. Nano-crystalline magnetic materials are metallic materials with a typical saturation flux density of 1.2 T. Their permeability can be tightly controlled between 20,000 and

S. O. Hwang et al. (Eds.): ICGHIT 2018, LNEE 502, pp. 31–36, 2019.
https://doi.org/10.1007/978-981-13-0311-1_6

100,000. These two parameters, saturation flux density and permeability, affect the size and performance of the couplers [5]. For instance nano-crystalline materials allow manufactures to build inductive couplers, which require magnetic cores up to several times smaller than similar cores made of ferrites [6].

In this study, we investigated the performance of an induction type PLC which can add communication equipment without changing the existing power distribution network in a ship. It is discussed how the magnetic characteristics of the toroidal core affect the coupling efficiency of the ICUs. From the computational simulation using the FEMM, it has been verified that the magnetic flux density of the nano-crystalline core is much higher than that of the ferrite. The relation between magnetic flux density and data transmission capacity was investigated. The nano-crystalline inductive PLC system showed better channel connectivity than the ferrite cores.

2 Experimental Results

Figure 1 shows the shape of the ferrite and nano-crystalline cores. Ferrites exhibit various properties depending on the kind and composition ratio of metal ions such as Fe, Mn, Ni, Mg, Zn, Cu, and Co. The basic structure of the magnetic inductor unit, first proposed by Mitsubishi Electric in 2005, is now one of the most important components in power electronic devices and is implemented as a high frequency filters, magnetic energy storages, EMC chokes, magnetic fault current limiters, etc. [7].

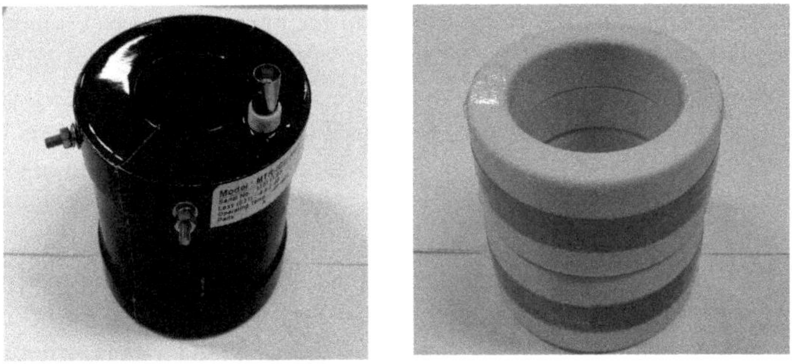

Fig. 1. The shape of toroidal cores. Ferrite core (left) and nano-crystalline core (right)

The first nono-composite soft magnetic alloy based on the Fe-Cu-M-Si-B alloy was obtained by Yoshizawa et al. [8]. A 20 mm thick ribbon was made of amorphous metal and rolled into a toroidal core. And then the core materials were annealed to produce the nano-crystalline magnetic material with an average grain size of 10 nm.

Figure 2 shows the frequency coupling characteristics for the three material cores. Nano-crystalline shows higher coupling efficiency in the operating frequency range than two ferrites. Mg-Zn ferrite is effective for blocking noise in the low frequency region of kHz. The upper limit frequencies are around 30 MHz and the insertion loss of

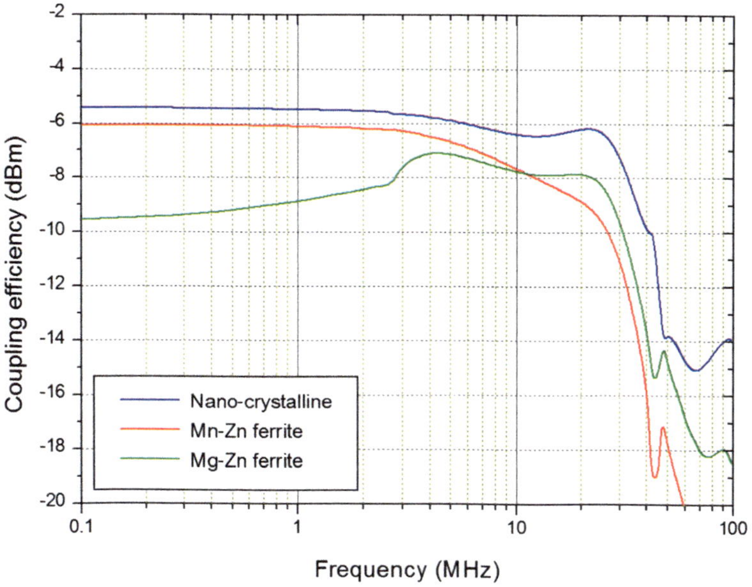

Fig. 2. Coupling efficiency for the different cores

the passband is about −5 dB on the average. The bandwidth almost coincides with the HD- PLC frequencies.

The ICUs contain a circular transformer consisting of magnetic cores and two wires sharing a magnetic field. By applying modulated signal to a primary winding, a current proportional to the coupling coefficient will be induced in a secondary winding. Therefore the core serves to couple the modulated signal from the primary winding to the secondary winding or vice versa. When the current (*I*) flows through the center of toroidal cores clamped over the powerline, the magnetic energy (W) accumulated in the core is expressed by Eq. (1).

$$\mathrm{W} = V_{core}\int_0^B H\,dB = \frac{1}{2}V_{core}\frac{B^2}{\mu} == \frac{V_e}{2}\mu_e\mu_0\left(\frac{N}{l_e}I\right)^2 \tag{1}$$

where B, V_{core}, μ, and l_e represent flux density, volume, permeability of the core, and magnetic path length, respectively. From Eq. (1), the size of the core and the kind of material are determined according to the magnitude of the current flowing in the power line.

To demonstrate the communication capacity of the distribution cable inside the vessel, an inductive PLC system was installed in two separate rooms as shown in Fig. 3. Experimental setup allows us to check whether an inductive PLC is available on the ship area network (SAN).

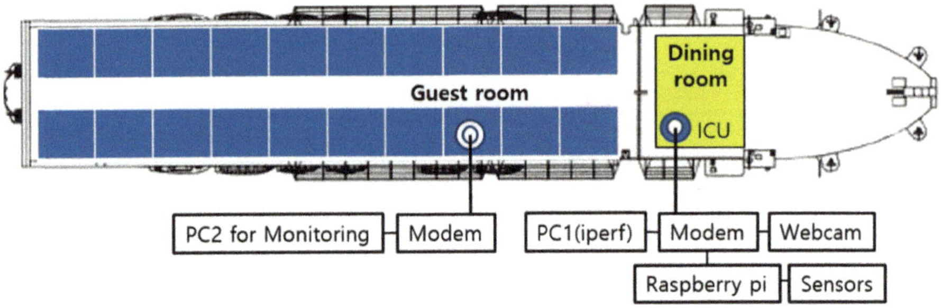

Fig. 3. Experimental configuration on the training ship.

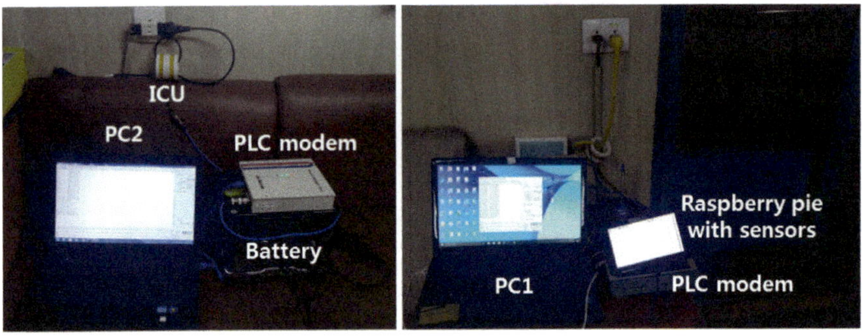

Fig. 4. Photograph of experimental setup. Guest room (left) and dining room (right).

We set the PLC modem installed in the guest room as the master and the other one installed in the dining room as the slave. Devices for data acquisition and transmission, such as a computer (PC1), webcam and raspberry pie, is connected to the slave modem as shown in Fig. 4.

The channel bandwidth can be monitored in the slaver iperf program as shown in Fig. 5. The average data rate measured 10 times in 30 s was more than 12.6 Mbps.

We also implemented a Raspberry pie sensor system that can transmit sensing data such as temperature and humidity by using a nano-crystalline PLC system and confirmed that the sensor data collected at the dining room can be transmitted to the guest room without errors. Real-time images captured with a webcam have also been successfully transferred to PC2. In the same experimental conditions, no communication was made in a PLC system using a ferrite inductive coupler. It can be seen that the magnetic flux density stored in the toroidal core affects the performance of an inductive PLC.

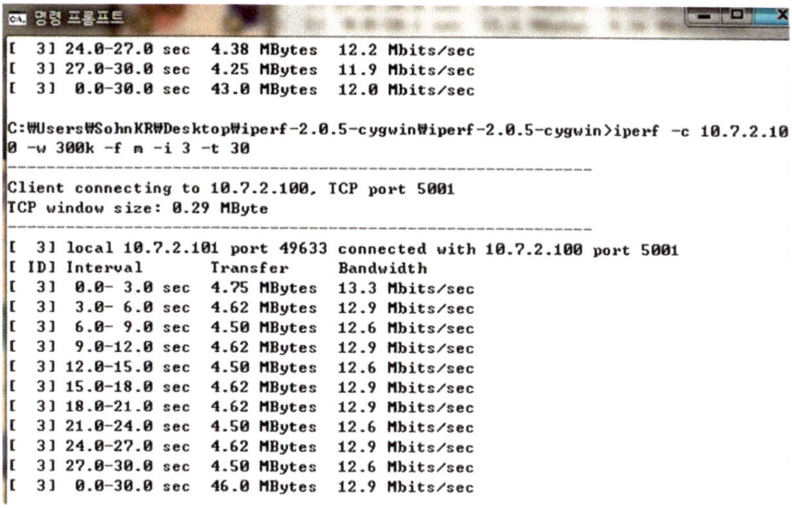

Fig. 5. Data transmission rate using a nano-crystalline inductive coupler

3 Conclusion

We demonstrated an inductive PLC using a nano-crystalline coupling unit that can be applied to the ship. Using a nano-crystalline inductive PLC, we achieved data rates of more than 12 Mbps between two separate rooms in the training ship and succeeded in transmitting images on the same embedded distribution. Because nano-crystalline cores have higher magnetic energies than ferrite cores, nano-crystalline inductive couplers are more suitable for inductive PLC applications in SANs.

Acknowledgement. This research was supported by Mid-career Researcher Program through the National Research Foundation of Korea (NRF) funded by the Ministry of Science and ICT (Grant No: NRF2017R1A2B4010993).

References

1. Sohn, K.R.: Inductive powerline communication using a soft magnetic core and its application. In: Asia Navigation Conference, China, vol. 1, pp. 1–6, November 2017
2. Jun, H.K., Kim, H.S., Jung, K.S., Sohn, K.R.: Development of bypass unit for ship area network based on legacy-line communication. J. Korean Soc. Marine Eng. **39**(3), 292–297 (2015). (in Korean)
3. Kim, H.S., Park, S.H., Kang, S.G.: Development of communication joint tools for implementing a legacy-line communication system in a train. J. Korea Inst. Inf. Commun. Eng. **19**(4), 877–887 (2015). (in Korean)
4. Kosonen, A., Ahola, J.: Comparison of signal coupling methods for power line communication between a motor and an inverter. IET Electr. Power Appl. **4**(6), 431–440 (2009)

5. Binkofski, J.: Influence of the properties of magnetic materials on the size and performance of PLC couplers. In: International Symposium on Power Line Communications and Its Applications, pp. 281–284 (2005)
6. Luciano, B.A., Albuquerque, J.M.C.: Nanocrystalline material in toroidal cores for current transformer: analytical study and computational simulations. Mater. Res. **8**(4), 395–400 (2005)
7. Murata, Y., Kimura, T.: Inductive coupling unit and bypass tool for power line communications. Mitsubishi Electric. Adv. **109**, 18–20 (2005)
8. Yoshizawa, Y., Oguma, S., Yamauchi, K.: New Fe-based soft magnetic alloys composed of ultrafine grain structure. J. Appl. Phys. **64**(1), 6044–6046 (1998)

Analysis of Hardware Requirements for IoT Teaching Aids

HyunChang Lee[1], KyuTae Lee[1(✉)], SeoIk Kang[1], KeunYoung Choi[1], WonSik Na[2], and SungYeol Kwon[3]

[1] Kongju National University, Cheonan 31080, Korea
{hclee,ktlee,auto}@kongju.ac.kr, valis0531@gmail.com
[2] Namseoul University, Cheonan 31020, Korea
winner@nsu.ac.kr
[3] Pukyong National University, Pusan 48513, Korea
sungyeol@pknu.ac.kr

Abstract. This paper proposes the requirements of teaching aids for IoT. It is necessary to protect the device from mistake by the user due to non-professional and educational device. In this paper, the case of failure was investigated and fault detection methods were proposed. The teaching aids with proposed protection circuit are not easily damaged by disturbance. Therefore, education can proceed more efficiently in terms of time and cost.

Keywords: IoT · Education tools · Teaching aids · Arduino

1 Introduction

By developing the computer technologies and communication technologies, devices for IoT (Internet of Things) technology have rapidly increased. The key is that IoT is a convergence of software and hardware technologies. Although education for IoT should be accompanied by software and hardware, current education system is conducted separately from them. It is difficult for the trainee to apply to the IoT field due to the lack of combining software with hardware technologies. Therefore, there is a need for other teaching methods, teaching materials and supporting aids for this purpose.

In recent years, teaching aids such as Arduino [1] and Raspberry Pi [2] is being used for IoT training. These are excellent tools for non-experts to conduct experiments, but not suitable for full-scale education. Lee et al. [3] analyzed the problems when these teaching aids were applied to education, and proposed the requirements for teaching aids. Reflecting the needs of IoT teaching aids, this paper analyzes the requirements for IoT teaching aids and proposed design to satisfy them.

2 Hardware Requirements for IoT Teaching Aids

The hardware of teaching aids is equipped with input ports, output ports, power supply etc. The case of failures and diagnosis of them are investigated by each part.

S. O. Hwang et al. (Eds.): ICGHIT 2018, LNEE 502, pp. 37–40, 2019.
https://doi.org/10.1007/978-981-13-0311-1_7

2.1 Output Port Protection Circuit

The output port could be shorted or the input signal could be applied it, because the trainee is not yet familiar with the hardware. Therefore, protection circuits are needed to prevent damage to the microcontroller of the device as shown in Fig. 1. If the input and output ports of the device are pre-allocated and fixed, a buffer circuit can be attached to isolate the output port from the microcontroller, and a current limit resistor can be used to protect the buffer output [4]. The resistor needs to be set to a proper value that does not affect the output current and large enough to protect the IC. In addition, a DIP (dual in-line package) type IC and a socket are used to replace the damaged buffer circuit.

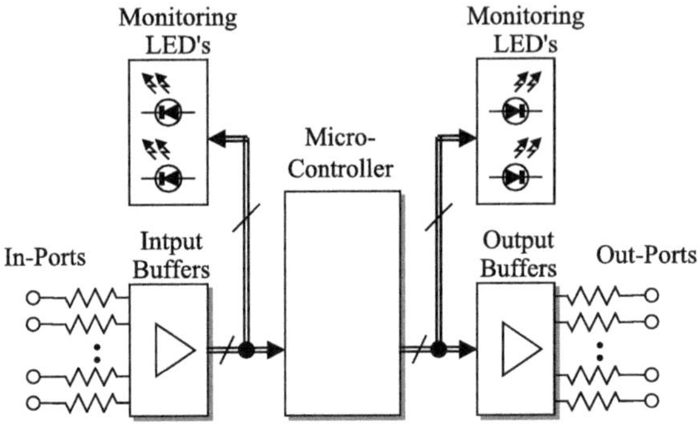

Fig. 1. Block diagram of input and output ports

2.2 Input Port Protection Circuit

There is no problem when the input port is short, but it is damaged by an over-voltage. In Fig. 1, the signal generated from the input port is transferred to the microcontroller via the buffer, and the input resistor is connected in series to protect it with a clipping diode which existed in the buffer [5]. It also needs to be set to a proper value that does not affect the input current and large enough to protect the IC. In addition, a DIP (dual in-line package) type IC and a socket are used to replace the damaged buffer circuit.

2.3 Port Monitoring LED

In order to check whether the circuit operates properly and to diagnose the fault of the device, the input port or the output port of the device is required to display an LED which indicates the status of the port as shown in Fig. 1. This indicator should be designed so as not to affect the logic.

2.4 Power Circuits and Protection Circuits

In general, the user makes the circuit with the power off in order to secure the circuit, and turns on the power after the circuit is completed. It is not desirable to turn on and off the power of the circuit in terms of life and stability. Therefore, a separate switch is proposed in the device as shown in Fig. 2. Although the switch is open, the power of the device is still being supplied and only the output port and power supplied to the user are turned off.

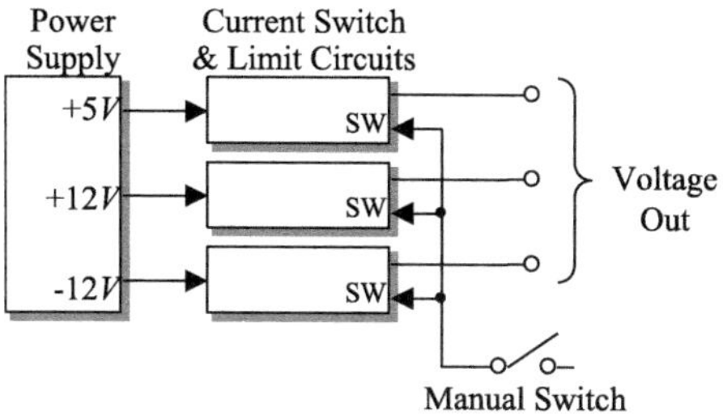

Fig. 2. Block diagram of the power supply unit

The power supply circuit needs to be supplied with three power sources: +5 *V* for logic circuit, and +12 *V*, −12 *V* for operational amplifier. Also, all the supplies should be supplied to the circuit simultaneously so that the circuit system operates stably. Since the power supply output is damaged by short-circuits, the current cut function should be installed to cut off the over-current. The diagnosis result is displayed to the user through the LED indicators.

2.5 Port Expansion Circuits

In general, an expansion connector is attached to a user to use an output or input port of the device. However, due to the large size of the extension connector (0.1 inch pitch) as shown in Fig. 3(a), Arduino cannot be used for general breadboard jumper wires, and a special wire should be used. In order to improve this problem, a small sized connector (2 mm pitch) is used for the breadboard as shown in Fig. 3(b). The 2 by 2 connector is more robust design, because the spare connector hole is available when the connector is damaged.

(a) 0.1 *inch* pitch (b) 2 *mm* pitch

Fig. 3. Connectors for circuit expansion

2.6 Expansion Slot

When a user studies software for hardware control, it is unnecessary to configure the circuit through port extension. For this purpose, other learning kits are considered to be able to insert hardware modules. However, there is a problem that the connector pins are bent and loose contact. In addition, it is difficult for the user to mount the connector pin accurately. In order to solve these problems, it is convenient to configure the hardware module as a card type and configure the card edge to fit in the slot. If the module is not properly inserted into the slot as shown in Fig. 4, it will cause a short circuit. Therefore, it is necessary to build the logic so that the power is supplied when the module is inserted correctly.

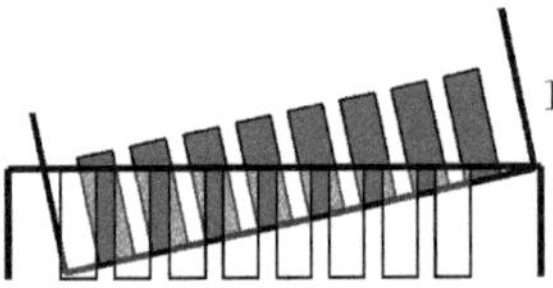

Fig. 4. Example of connection failure in module and slot

3 Conclusion

In this paper, the hardware requirements of IoT teaching aids were proposed. This device is not a professional device but educational device and needs to be protected from misuse by the user. The possible faults of each element of the device were investigated and protection methods were proposed. The educational device with proposed protection circuit is not easily damaged. Therefore, education can proceed more efficiently in terms of time and cost.

References

1. Blum, J.: Exploring Arduino: Tools and Techniques for Engineering Wizardry. Wiley, New York (2013)
2. Philbin, C.A.: Adventures in Raspberry Pi. Wiley, New York (2014)
3. Lee, K.-T., Lee, H.-C., Lee, J.-E.: Analysis about construction conditions of the auxiliary hardware teaching materials for software education. In: 27th Conference of Korea Software Assessment and Valuation Society, pp. 367–368 (2017)
4. Vahid, F., Givargis, T.: Embedded System Design - A Unified Hardware/Software Introduction. Wiley, Hoboken (2002)
5. Morris Mano, M., Ciletti, M.D.: Digital Design, 4th edn. Pearson, México (2003)

Location Estimation Technique in Bluetooth Beacon Based Indoor Positioning Systems

Jeong Hyun Yoon, Inah Chung, and Ye Hoon Lee(✉)

Seoul National University of Science and Technology, Seoul 01811, Korea
y.lee@snut.ac.kr

Abstract. Recently low power based Bluetooth beacons have been developed, and the received signal strength indication (RSSI) based distance measurement technique of beacon signal has been actively studied. In order to calculate the accurate position information, reliability of the received RSSI value and an efficient calculation method are needed. However, it is considered that the calculation method of the conventional triangular algorithm which is currently popular is difficult to recognize the accurate position information due to the accuracy limitation. Therefore, in this paper, we propose a novel weighting algorithm that compensates the limits of the triangulation algorithm. When we compare the accuracy with the actual measured data, we can confirm that the error rate of the location information obtained using the proposed weighting algorithm method is lower than that of the conventional triangulation algorithm.

Keywords: Bluetooth beacon · Indoor positioning · RSSI

1 Introduction

Recently, various studies have been conducted on indoor location measurement to enhance convenience of consumers. Since the Global Positioning System (GPS), which is used to measure the existing location, is a service limited to the outdoor area, it is required to construct a positioning system using a radio transmitter capable of receiving indoors. In a positioning system based on a signal of a wireless device, when the radio propagation is not a free space, the propagation path between the transceivers is blocked by objects, or when there is an irregular surface, the accuracy of the signal due to path attenuation occurs. Therefore, it is not easy to obtain accurate indoor position by receiving wireless signals.

In this paper, we investigate the accuracy of the received signal strength indication (RSSI) received through a signal transmitted from a Bluetooth beacon and calculated the indoor position using the RSSI value. The triangulation algorithm, which was mainly used in the past, shows a tendency that the error rate to the position is considerably large due to the influence of the RSSI value which is not constant [2]. Therefore, we consider a new location estimation algorithm based on the ratio of signal strength received from each beacon and show the validity and superiority of the proposed algorithm based on the indoor measurement data.

S. O. Hwang et al. (Eds.): ICGHIT 2018, LNEE 502, pp. 41–44, 2019.
https://doi.org/10.1007/978-981-13-0311-1_8

2 System Model

In this paper, we compare the triangulation algorithm with the proposed weighting algorithm to find out the improvement of the position estimation error according to the difference of the position estimation methods. Since there is a difference in accuracy depending on the type and characteristics of the space, our experiments were done in an open space and an closed space by using beacon signals as a Bluetooth module and calculating them using MATLAB.

2.1 Triangulation Algorithm

The distance equations required in the triangulation algorithm used in the experiments were modeled by Texas Instruments for the Chipcon CC2420 ratio

$$\mathrm{RSSI} = -((10\mathrm{n}\log d) - A) \tag{1}$$

where n is the path loss parameter for a particular space and A is the measured RSSI dBm value when the distance between the transceivers is 1 m [1] (Fig. 1).

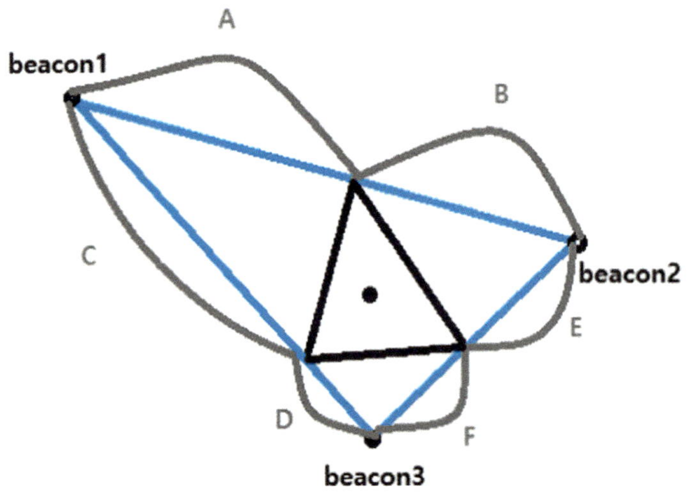

Fig. 1. Calculation method of the proposed weighting algorithm

2.2 Proposed Weighting Algorithm

The main difficulties of the triangulation method occurs when the value of d is too large/small depending on the RSSI value or there is no intersection of the circles. In order to overcome such problems, we propose a novel location estimation algorithm called as *weighting algorithm*. Unlike the triangulation algorithm, the proposed weighting algorithm exploits the ratio of received RSSI values without using a circle. The proposed method estimates the position by internally dividing each side of a large external triangle connecting each beacon and obtaining the end point and then calculating the center of gravity of the small triangle inside.

3 Experiments and Results

In our experiments, prior to using the triangulation algorithm, we first set the n and A values for triangulation. First, we measure the value of A, which represents the distance between the transceivers is 1 m. As a result of the incorrect use of the RSSI value for position location [1], ten RSSI values at a distance of 1 m for each beacon were received and the average of the RSSI values was used as the actual RSSI value. As a result, in the closed space, beacons 1 to 6 are −58.1, −61.2, −59.7, −55.54, −59.66, and −57.41 dBm, respectively, 50.8 dBm is measured. The value of n is calculated from Eq. (1) based on the value of A. The result value of A is 1.79 in the closed space and 1.71 in the open space. As in the calculation of the A, the RSSI value takes several RSSI measurement values and uses the average [2].

Prior to the experiment, beacons are placed at the corners at intervals of 7 m in the experimental space to reduce the maximum positioning error [3], and the MATLAB code is implemented to receive the RSSI value using the Bluetooth module. Measurements are made in units of 1 m, and the positions are measured using both the triangulation algorithm and the proposed weighting algorithm for each measurement. Additional attention is paid to the height of the beacon and receiver during the experiment. In Eq. (1), the difference of the RSSI value according to the height is not calculated, so that the height of the beacon and the receiving part can be maximized as much as possible.

Experiments are carried out in an enclosed space and an open space (lobby) to analyze the error rates according to different spaces. Tables 1 and 2 show the averages of the corners and the internal errors because the beacons are placed at the corners, which makes it difficult for the Bluetooth module to receive the direct wave of the beacon on the same wall. Therefore, when selecting a beacon close to the receiver, a beacon at the opposite corner, that is, a beacon having a high probability of receiving a straight line signal is often selected. Since the resulting edge error is relatively larger than the internal error, it is analyzed separately for accurate comparison.

Table 1. Experimental results in an enclosed space.

Enclosed space	Average at corner (m)	Average at internal space (m)
Weighting algorithm	2.7537	1.8637
Triangulation algorithm	2.9545	2.5558

In the closed space (Table 1), when the internal averages are compared, it is found that the weighting algorithm yields smaller error by about 0.7 m than the triangulation algorithm. In particular, in the open space (Table 2), it is confirmed that the error rate is relatively high because the variation range of the beacon with the RSSI value is large.

Table 2. Experimental results in an open space.

Open space	Average at corner (m)	Average at internal space (m)
Weighting algorithm	3.5531	2.6220
Triangulation algorithm	7.7917	3.6574

In this experiment, it is shown that the positioning error of the weighting algorithm is about 1 m smaller than the triangulation algorithm.

4 Conclusions

In this paper, we have examined the indoor location problem to improve the existing algorithm that calculates the position using triangulation by receiving RSSI signal of beacon. Since the RSSI signal is unstable due to irregularities of obstacles and reflectors in the experimental space, accurate calculation cannot be made. Therefore, the weighting algorithm is proposed to exploit the ratio of received RSSI values instead of the signal intensity. The experimental results show that the proposed weighting algorithm is superior to the triangulation algorithm in terms of overall average of location error.

Acknowledgments. This work was partly supported by Institute for Information & communications Technology Promotion (IITP) grant funded by the Korea government (MSIT) (No. 2016-0-00144) and by Basic Science Research Program through the National Research Foundation of Korea (NRF) funded by the Ministry of Education (No. 2017R1D1A1B03035522).

References

1. Goldoni, E., Savioli, A., Risi, M., Gamba, P.: Experimental analysis of RSSI-based indoor localization with IEEE 802.15.4. In: Proceedings on European Wireless (EW) Conference, pp. 71–77. IEEE, Lucca (2010)
2. Dong, Q., Dargie, W.: Evaluation of the reliability of RSSI for indoor localization. In: Proceeding on International Conference on Wireless Communications in Unusual and Confined Areas (ICWCUCA). IEEE, Clermont-Ferrand (2012)
3. Ahn, H., Thuy, T.V., Lee, Y.H.: Bluetooth beacon planning considering position estimation accuracy in small and isolated indoor environment. J. Korean Inst. Commun. Inf. Sci. **40**(7), 1307–1312 (2015)

Fog Agents in Distributed Multi Agent System for Autonomous Energy Management

Ui-Kyun Na, Young-Gon Kim, Jae-Seong Jo, and Eun-Kyu Lee(✉)

Incheon National University, Incheon, Republic of Korea
eklee@inu.ac.kr

Abstract. Fog Computing is an emerging paradigm that distributes computing and networking services closer to where user data is generated. The Fog presents a new architecture that "takes processing to the data". The fog will resolve in the Internet of Things such as bandwidth and cost constraints and performing data fusion, etc. Considering the "distributed" property, this paper proposes to exploit the multi agent system technology for the development of the fog system. To demonstrate its feasibility, we adopt an open source platform for a multi agent system, VOLTTRON. Then, four types of fog agents are developed that collaborate to conduct building energy management tasks in an autonomous manner. The agents can reduce the energy waste in a building by managing the brightness of lights, power status of the HVAC systems, and various energy loads. They also receive weather data and price data from Korea Meteorological Administration and Korea Power Exchange and present the data on the web. This helps user to recognize the information easier.

Keywords: Energy management system · Fog computing · Multi agent system
Internet of things · Smart grid

1 Introduction

A new paradigm of fog computing and networking is emerging to resolve many challenges that current cloud computing confronts – e.g., processing data in real time, creating location-aware contexts, and maximizing the efficiency of last mile wireless communications. To overcome the limitations, portions of computing and networking capability of the cloud move down to the network edge and form a local computing environment, a *fog*. The fog presents a new architecture that "takes processing to the data" while the cloud "takes data to the processing." Edge devices and mobile devices are inter-connected together within a local fog network, and they collaboratively carry out storage, data processing, networking and control tasks. The fog will resolve many problems in the Internet of Things (IoT). For instance, fog services can overcome the bandwidth and cost constraints for long-haul communications; performing data fusion and streaming analytics in real time; managing a large volume of devices and cyber-physical interactions; coping with network reliability and resiliency; and securing untethered, resource constrained end devices. With the potential benefits, the fog enjoys a variety of applications including home and building energy management system. Considering the "distributed" property in the fog, this paper proposes to exploit

S. O. Hwang et al. (Eds.): ICGHIT 2018, LNEE 502, pp. 45–50, 2019.
https://doi.org/10.1007/978-981-13-0311-1_9

a distributed multi agent system technology for the development of the fog system. To show the feasibility, we develop fog agents for autonomous energy management, running on top of open source multi agent platform.

2 Background

2.1 Energy Management System

Energy Management System (EMS) refers to a computer aided instrument System used by an operator of an electrical utility grid to monitor, control, and optimize the performance of the generation and/or transmission system. It is also used by individual, commercial entities such as retail stores or restaurants to monitor, measure, and control loads on buildings, and can be used to centrally control devices such as HVAC units and lighting systems. Building Energy Management System (BEMS) is a type of EMS system specialized for buildings. It often includes additional functions of energy management and environmental management, management and analysis of building energy facilities, and integrated management of Building Automation System (BAS) central system. By applying BEMS, it is possible to control the indoor environment and carbon emission, and reduce energy consumption by an average of 5–15% [1].

2.2 Agent Based Platform

An agent is a computer system that is situated in some environment, and that is capable of autonomous action in this environment to meet its design objectives [2]. Agents take sensory input from their environment, produce output actions that affect it. Sycara et al. [3] characterized Multi-Agent Systems (MAS) as distributed systems with two or more agents. MAS, featuring the implementation and utilization of multiple distributed intelligent agents, shares many common characteristics such as being adaptive, self-aware and semi-autonomous or autonomous. The most outstanding advantages that MAS embraces are that they can respond to the external environment rapidly, and can provide timely solutions based on distributed control with or without human intervention. Agent based technology has recently shed light on home and building energy management [4, 5].

Building a MAS requires an agent development environment that supports at least some stages of the MAS conceptual design process. Foundation for Intelligent Physical Agents (FIPA) specifications define rules for the existence, operation and management of generic agents that can be combined to make complex systems [6]. FIPA also allows for the abstraction of the agent development platform from the language used to program the agent, its environment, domain, data and external devices. As of today, there are several open source agent development platforms most often used by researchers for developing MAS for micro-grid control applications. This paper gives a brief review of three most popular ones: JADE, ZEUS, and VOLTTRON.

2.3 JADE, ZEUS, and VOLTTRON

Java Agent Development Environment (JADE) [7] is an open source agent development software framework for building FIPA compliant MAS. JADE provides tools for building distributable agents across multiple hosts while supporting parallel and concurrent agent activities. It is fully implemented in JAVA programming language, and thus developers can easily build a FIPA-compliant multi-agent system with their set of Java classes. The disadvantage is limited support on resource management and security that are important requirements for our platform.

ZEUS [8] is a FIPA compliant open source agent development platform implemented in Java programming language. ZEUS allows the modeling of agent roles using combination of class diagrams and predefined roles common in most management systems. This design environment restricts the need for developing new formalism, thereby making MAS development more accessible to a wider audience. But, new developers might face some challenges in creating new applications using the ZEUS platform owing to weak documentation.

VOLTTRON [9] is a distributed agent execution framework designed by a Pacific Northwest National Laboratory (PNNL) specifically for use in electrical power systems. The open source and modular platform is designed to be able to run on small-form factor computers, capable of interfacing with hardware devices, maintain security, manage platform resources, and service for applications. The VOLTTRON platform enables the deployment of intelligent sensors and controllers in residential/commercial buildings and the smart grid. Peer-to-Peer communications among distributed agents are established through a central *Message Bus* in the form of topics and subtopics (e.g., "topic/subtopic/subtopic/" = "weather/location/Temperature/"). VOLTTRON also provides driver support for most Modbus and BACnet devices.

JADE is best suited for advance developers building stable and scalable MAS control for micro-grids. ZEUS could be best employed by new developers for rapid prototyping and testing of MAS concepts for micro-grid control. VOLTTRON is best employed by facilities and building and managers for managing sensor and instrumentation data. In this paper, VOLTTRON is chosen to be an agent development platform for the proposed platform. While VOLTTRON is relatively new and continually evolving platform, its three properties (the open source development platform, broad driver support, and programming language independence) allows developers to prototype and demonstrate MAS control of grid devices even without access to testing hardware.

3 Development of Fog Agents for Autonomous Energy Management

This paper proposes several fog agents that collaborate for autonomous energy controls. They communicate over a secured message bus. The agents are developed on top of the VOLTTRON platform, and Fig. 1 shows the overall system architecture.

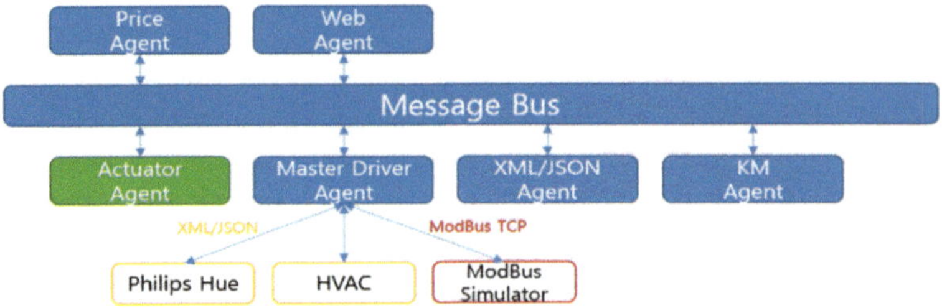

Fig. 1. A system architecture of distributed multi agent system on VOLTTRON platform for energy management, it includes a list of fog agents

3.1 Physical Device Agent

A physical device agent is responsible for connecting to physical devices whose functions are related to generation, consumption, and storage of power. These include office appliances like lights and plugs as well as energy facilities in buildings like solar panel and battery. Any sensors and actuators that can help energy management are also considered as physical devices. As an instantiation of the physical device agent, an XML/JSON Agent communicates with light (Philips Hue) and Heating, ventilation, and air conditioning (HVAC) via a thermostat (CT-50). Figure 2 depicts a list of physical devices deployed in our testbed. It can control the devices as well as to read status data from them. For experiments, we have also setup a ModBus Simulator that mimics operations of many physical devices communicating with the agent via ModBus communications, which is shown in Fig. 3.

Fig. 2. Pictures of three devices: Thermostat, LED light, and smart plug

3.2 Price Agent and Weather Agent

A price agent communicates with an external service that provides power price in real time. Different countries have their own standards for delivering real-time prices; and some may not have policies for real-time prices; and some may not have policies for real time prices yet. In our implementation, the price agent connects to web APIs provided by Korea Power Exchange (KPX) and obtains System Marginal Price (SMP) [10]. The price data is then shared with other fog agents with appropriate topic names via the message bus.

A weather agent operates in a similar way. Instead of price data, it obtains weather data via web API provided by Korea Meteorological Administration (KMA) [11]. The

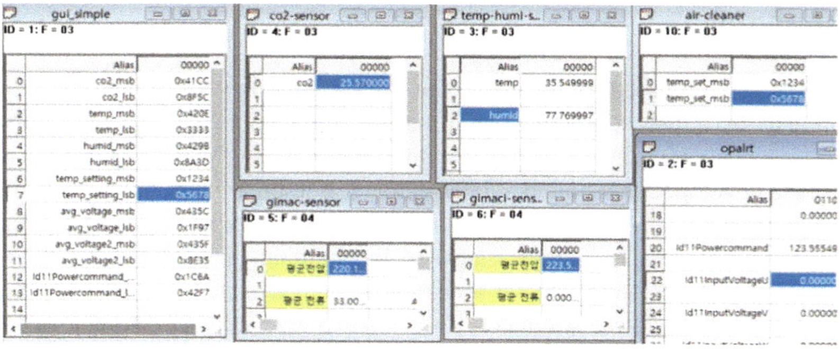

Fig. 3. Our testbed also setups a ModBus simulator that allows experiment for scalability.

data is shared with other agents via the message bus. The data is retrieved in terms of GPS coordinates and the obtained one contains 13 different types of details including 3-hour-ahead temperature, humidity, wind, rain, cloud, min/max temperature of the day, etc. A 3-day forecast is also included in the weather data.

3.3 UI Agent

A User Interface (UI) agent is responsible for interacting with end users (human beings) regarding energy information. It informs users of energy status as well as accepts control commands from them. To implement two functions, we develop a simple web agent that communicates with the other agents (price agent, weather agent, and physical device agent) via the message bus. Upon receiving data, the web agent displays the data on the web that is accessible via a web browser. Figure 4 captures the web agent displaying the real-time price and weather information on the web.

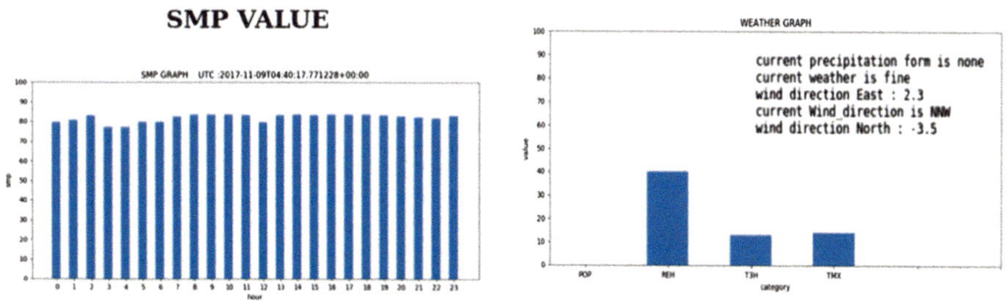

Displaying a real time-price and weather data

Fig. 4. The UI agent shows two different types of energy-related data on the web - the real-time price and weather data

4 Conclusion

This paper has proposed to leverage the multi-agent system technology for the development of a fog computing environment. To demonstrate its feasibility, we adopted an open source platform for a multi agent system, VOLTTRON. Then, fog agents were developed that collaborate to conduct building energy management tasks in an autonomous manner. The agents can reduce the energy waste in the building by realizing the brightness and power status of the lighting in the building, the power status of the HVAC system, the set temperature and the fan mode in real time and implementing it through the simulator. They also receive weather data and price data and present it on the web to provide information to the user. Advancing the agents and complete development of the fog system for BEMS are integral portion of our future works.

Acknowledgement. This work was supported by the National Research Foundation of Korea (NRF) grant funded by the Ministry of Science, ICT & Future Planning (No. 2016R1C1B1 016084). Eun-Kyu Lee is the corresponding author.

References

1. BEMS Homepage, http://www.bems.or.kr/inc.php?id=sub04_ct01. Last accessed 05 Jan 2018
2. Wooldridge, M., Jennings, N.: Intelligent agents: theory and practice. Knowl. Eng. Rev. **10**(02), 115–152 (1995)
3. Sycara, K.: Multi agent systems. AI Mag. **19**(2), 79–92 (1998)
4. Rui, Y., Lingfeng, W.: Multi-agent based energy and comfort management in a building environment considering behaviors of occupants. In: IEEE Power and Energy Society General Meeting, pp. 1–7. IEEE (2012)
5. Asare-Bediako, B., Kling, W., Ribeiro, P.: Multi-agent system architecture for smart home energy management and optimization. In: IEEE Innovative Smart Grid Technologies Europe (ISGT EUROPE), pp. 1–5. IEEE (2013)
6. Foundation for Intelligent Physical Agents Homepage. http://www.fipa.org/. Last accessed 10 Dec 2017
7. Bellifemine, F., Poggi, A., Rimassa, G.: JADE: a FIPA2000 compliant agent development environment. In: Proceedings of International Conference on Autonomous Agents, pp. 216–217. ACM (2001)
8. Nwana, H., Ndumu, D., Lee, L., Collis, J.: Zeus: a toolkit and approach for building distributed multi-agent systems. In: ACM Annual Conference on Autonomous Agents, pp. 360–361. ACM (1999)
9. Akyol, B., Haack, J., Ciraci, S., Carpenter, B., Vlachopoulou, M., Tews, C.: VOLTTRON: an agent execution platform for the electric power system. In: International Workshop on Agent Technologies for Energy Systems, Spain (2012)
10. Korea Power Exchange Homepage. https://www.kpx.or.kr/eng/index.do. Last accessed 05 Dec 2017
11. Weather data, Korea Meterological Administration Homepage. http://www.weather.go.kr/weather/main.jsp. Last accessed 10 Dec 2017

Building Entry Loss (BEL) Characteristics for Incident Angles and Measurement Locations from 3.5 to 24 GHz

Young Chul Lee[1(✉)], Soon-Soo Oh[2], Jae-Won Choi[2], Hwa Choon Lee[2], Jong-Hyuk Lim[3], Dae-Hwan Yoon[3], Sung Won Park[3], and Byung-Lok Cho[4]

[1] Mokpo National Maritime University (MMU), Mokpo, Republic of Korea
rfleeyc@gmail.com
[2] Chosun University, Gwangju, Republic of Korea
[3] National Radio Research Agency, Naju, Republic of Korea
[4] Sunchon National University, Suncheon, Republic of Korea

Abstract. An effect of a building entry loss (BEL) on incident angles has been investigated in a traditional office building from 3.5 to 24 GHz band. The BEL was measured at three different positions in the window, the office center, and the corridor, moving a receiver (Rx) from 1 m away from the window (W) to the inside on the 6th (6F) and 9th (9F) floor of the building. As the Rx location moves from the window (W) to O and C inside of the building, the BEL increases and its slope for the azimuth angle decrease on the 6F, where the signal is directly incident. This tendency was not observed in 9F with an elevation angle of 12.8°. These results reveal that there is a breakpoint and propagation mechanism is different at each Rx location. For cumulative distribution function (CDF) of the BEL, as the frequency increases to 3.5, 6, and 10 GHz, the BEL increases. However, it decreases at 24 GHz at all Rx locations and all elevation angles because of frequency dependency of the input impedance of the air dielectric interface in the double layered-glass window. As the elevation angle increases from 0 to 12.8° in the window, the BEL increases from 4.2 to 11.7 dB at 50 percentile level of the CDF.

Keywords: Building entry loss (BEL) · Penetration loss · Outdoor-to-indoor propagation

1 Introduction

Emergence of various wireless communication services such as mobile, WiFi, ZigBee, and internet of thing (IoT) leads to expand available frequencies. Because radio networks have to be carefully planned and optimized, more precise and more diversified radio propagation models are needed over many bands or at high frequencies. Recently, outdoor-to-indoor (O2I) path loss and building entry loss (BEL) are emerging as important issues, because of the large loss contribution, which must be considered when designing radio link networks [1].

S. O. Hwang et al. (Eds.): ICGHIT 2018, LNEE 502, pp. 51–56, 2019.
https://doi.org/10.1007/978-981-13-0311-1_10

Several empirical models and measurement analysis have been investigated under the names of O2I path loss, O2I penetration loss, and O2I propagation loss with different definitions [2–4]. Therefore, these research results are not easy to be applied. In addition, since there are many variables to be considered, such as the environment, the location of a transmitter (Tx) and receiver (Rx), building materials, and its structure, many difficulties are encountered in implementing accurate and standardized models. Moreover, some of the measurement results show clearly contradictory observations [5–8]. Obstruction inside the Fresnel zone or frequency dependency due to a layered glass in a window is one of the main causes [5]. Some study demonstrated that the frequency dependence can be improved by considering multiple reflected and diffracted waves [8]. However, the difference is not clear at millimeter wave. In general, although the BEL increases with increasing incident angle [7, 8] or receiving distance [9], frequency dependency was still unclear. That means that propagation mechanism is different near the window and inside the building. Therefore, various variables and complex propagation mechanisms make it difficult to analyze the measurement results.

In this paper, the characteristics of the BEL have been measured with varying elevation and azimuth angle while changing the position of the Rx inside the building from 3.5 to 24 GHz.

2 Measurement Environment and Scenario

The definitions of BEL and their measurement methods are well documented in the ITU-R document [10–12]. In this work, the O2I BEL measurement was performed in an environment shown in Fig. 1. This traditional office building is with reinforced concrete shear wall and double glazing window without a metal coating. Its 6 mm thick two glasses are separated by the air gap of 12 mm. A thickness of a bearing wall is 35–38 cm. The interior wall is being equipped with a plaster board and foam polyethylene sheet and its thickness is around 25 cm. An outdoor Tx antenna in the front of the building is fixed on a road. In order to investigate the effect of the BEL on an incident angle of azimuth and elevation, an indoor Rx is located in 6F and a received power was measured while moving the Rx as shown in Fig. 1(a). The height of the Rx antenna in the 6F is the same as the Tx antenna. Measurements were repeated at the 9th floor (9F) with different elevation angle. The height difference between 6F and 9F is 10.5 m and the corresponding elevation angle is 12.8°. There is a six-floor deep valley between the building and road as shown in Fig. 1(c).

Figure 2 shows details of the BEL measurement scenario. One fixed outdoor Tx and 28 different Rx points on each floor of the building were selected as shown in Fig. 2(a). Seven reference locations were selected with different incidence angles of azimuth (α), which are from 19.0 to 60.1°. Reference power was measured at a distance of 1.6 m from the exterior wall of the building. By using one reference, the BEL was characterized at three different locations inside the building with the distance of 1 and 5 m from the window in the exterior wall and in the middle of the corridor and their locations are named window (W), office (O), and corridor (C), respectively, as shown in Fig. 2(b). This measurement was repeated in the same manner in the 9F with the incident angle of elevation of 12.8° as shown in Fig. 2(c).

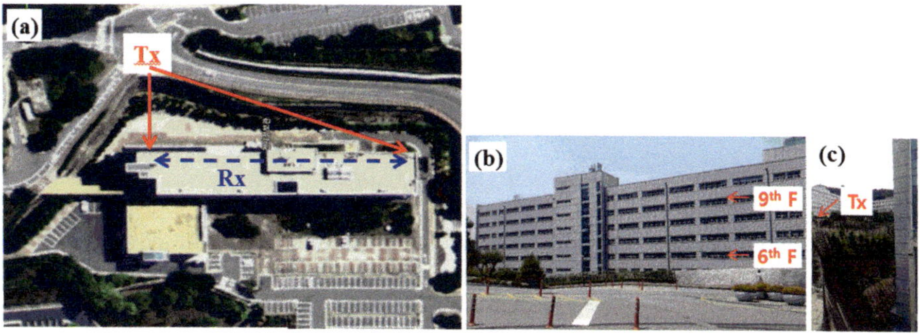

Fig. 1. Photographs of the building for the BEL measurement on Chosun university campus (Gwangju, in Korea). (a) Tx position on the road, and Rx movement route in the building, (b) the external façade of the building, and (c) a valley between the Tx and the building

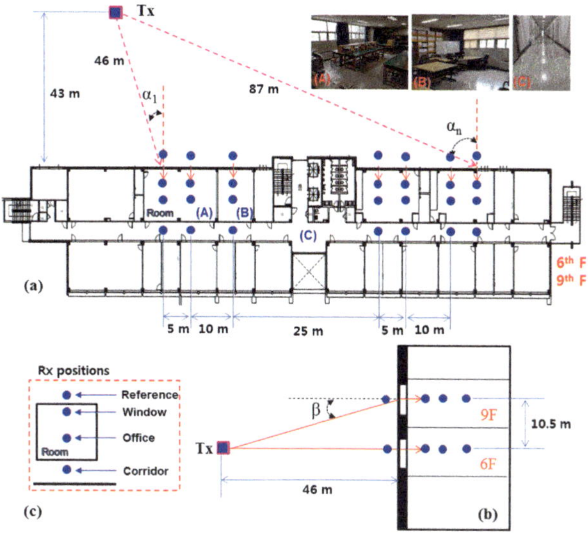

Fig. 2. Details of the BEL measurement scenario. (a) Reference and indoor Rx positions in the 6F and 9F floor in the building, the incident angle of azimuth (α), and photographs of university laboratories (a, and b) and corridor (c), (b) the incident angle of elevation (β) for the floor height, and (c) name for Rx points; window (W), office (O), and corridor (C).

Measurements were taken by using continuous wave (CW) of 3.5, 6, 10, and 24 GHz. The antennas used for both the outdoor Tx and indoor Rx were omni-directional ones. The Tx antenna was mounted on a pole on the roadside whose height is 2.5 m above the ground. The Rx antenna was fixed on a handcart inside the building and its height was 1.5 m above the floor. The Tx consisted of a signal generator (SG) and a power amplifier connected with the antenna. The Rx was composed of the identical antenna, low-noise amplifier and spectrum analyzer.

3 Measured Results and Analysis

The BEL in this campaign has been calculated by using definition in Recommendation ITU-R P.2040-1 [12] as:

$$BEL = P_{ref} - P_{indoor}[dB]$$

where BEL is the building entry loss, P_{ref} is the spatial median of the power received outside the illuminated face of a building, and P_{indoor} is the received power at the indoor locations.

The median value of the BEL for the incident angle of each frequency is calculated. Figure 3 presents the BEL characteristics for the incident angle at each measurement location. The BEL tends to increase with increasing the incident angle of azimuth at all locations. The BEL in 9F (β = 12.8°) increases, compared to the BEL in 6F (β = 0°). As the location moves from window (W) to office (O) to corridor (C), the BEL tends to increase. When the incident angle of azimuth increases beyond 52°, the BEL characteristics change significantly irrespective of the frequency and location. It is obvious that only a small part of the incident wave enters into the building and there is a complicated path of reflection and diffraction. However, the frequency dependence of the incident angle of azimuth is not clear and the BEL at 24 GHz is low. In general, the input impedance of the air dielectric interface in the double layered-glass window depends on frequency [5].

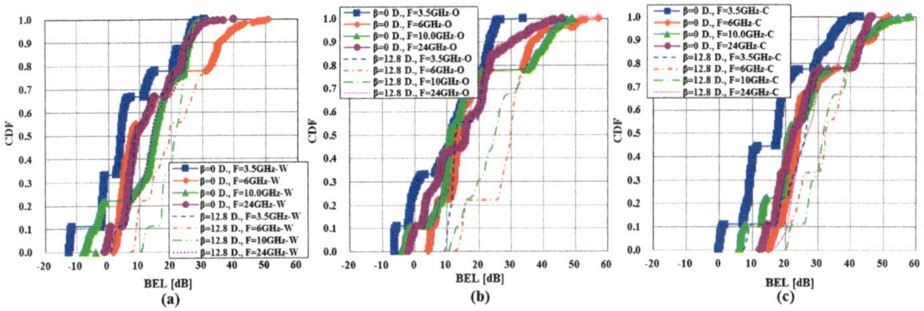

Fig. 3. Incident angle dependency of the BEL for each Rx location (a) window (W), (b) office (O), and (c) corridor (C) [D: degree].

By using a linear regression analysis, the linear regression lines at 3.5 GHz in each location are shown in Fig. 3. The BEL slope for the incident angle of azimuth in each location is extracted. The results are summarized in Table 1. As the measuring position moves from the window to the inside of the building, the BEL increases and its slope decrease on the 6F (β = 0°). This tendency was not observed in 9F (β = 12.8°). A consistent change in the BEL slope means that a breakpoint is present and propagation mechanism is different for each location. In the case of the 9F with a large elevation angle, a part of the radio wave illuminated on the exterior wall enters into the building and it seems to be influenced by the complex clutters inside. For the window location, the BEL per an incident azimuth angle is 0.335 dB/degree at 3.5 GHz and the highest value of 0.413 dB/degree is observed at 10 GHz. For other frequencies, the trends are similar. However, there was no frequency dependence of the slope.

Table 1. BEL slope for the incident angle of azimuth [dB/degree]

Location	Window (W)		Office (O)		Corridor (C)	
β [degree]	0	12.8	0	12.8	0	12.8
3.5 GHz	0.335	0.380	0.382	0.202	−0.032	0.218
6.0 GHz	0.180	0.281	0.069	0.344	−0.184	0.271
10 GHz	0.413	0.188	0.241	0.392	0.295	0.159
24 GHz	0.186	−0.075	0.135	−0.207	0.039	0.148

Figure 4 shows the cumulative distribution function (CDF) of the measured BEL for each Rx location. On the 6F, the minimum point of the BEL increases by 11.2 dB when the Rx location moves from the W to O and the C. The CDF distribution on 6F is wide, compared to that on 9F. At 50 percentile level of CDF, as the frequency increases to 3.5, 6, 10 GHz, the BEL increases, but it decreases at 24 GHz at environments (W, O, and C) and β. In the W and C, as the elevation angle increased from 0 to 12.8°, the BEL increase range is from 4.2 to 11.7 dB and also its fluctuations is not very large. However, in the office, as the elevation angle increases, the BEL fluctuates significantly. In the case of W and C, the structure is simple and obstacles in the propagation path are a few, compared to the O.

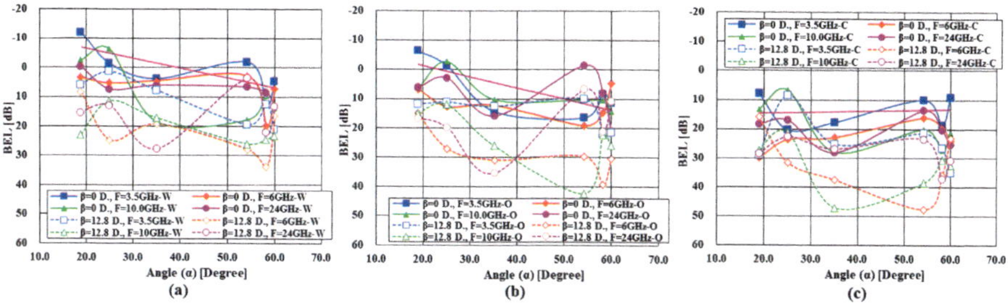

Fig. 4. CDF of the BEL for each Rx position. (a) window (W), (b) office (O), and (c) corridor (C).

4 Conclusion

In this work, the BEL of the traditional office building has been measured and analyzed from 3.5 to 24 GHz bands. Its characteristics for the incident angles were analyzed at three different locations in the 6F and 9F with elevation angles of 0 and 12.8°, respectively. The BEL tends to increase with increasing the azimuth and elevation angle at all locations. As the Rx location moves from the window (W) to office (O) and corridor (C), the BEL increases, but its BEL slope for the azimuth angle decreases. These results reveal that a breakpoint is present and the propagation mechanism is different for each location. Especially, the BEL slope for the azimuth angle at 24 GHz is 0.186, 0.135, and 0.039 dB/degrees for W, O and C, respectively. At 50 percentile level of CDF, as the frequency increases from 3.5 GHz to 6 and 10 GHz, the BEL increases, but it also decreases at 24 GHz at all measurement locations and β.

Acknowledgment. This research was supported by a grant of the research and development fund from the National Radio Research Agency (RRA) in Korea, 2017 and supported by Basic Science Research Program through the National Research Foundation of Korea (NRF) funded by the Ministry of Education (2017R1D1A3B03036543).

References

1. Okamoto, H., Kitao, K., Ichitsubo, S.: Outdoor-to-indoor propagation loss prediction in 800-MHz to 8-GHz band for an urban area. IEEE Trans. Veh. Technol. **58**(3), 1059–1067 (2009)
2. Castro, G., et al.: Outdoor-to-indoor empirical path loss models: analysis for pico and femto cells in street canyons. IEEE Wirel. Commun. Lett. **6**(4), 542–545 (2017)
3. Kim, M.-D., Liang, J., Lee, J., Park, J., Park, B.: Path loss measurements and modeling for indoor office scenario at 28 and 38 GHz. In: International Symposium on Antennas and Propagation (ISAP) (2016)
4. Imai, T., Kitao, K., Tran, N., Omaki, N., Okumura, Y., Nishimori, K.: Outdoor-to-Indoor path loss modeling for 0.8 to 37 GHz band. In: European Conference on Antennas and Propagation (EuCAP), pp. 1–4 (2016)
5. Stavrou, S., Saunders, S.R.: Factors influencing outdoor to indoor radio wave propagation. In: International Conference on Antennas and Propagation (ICAP), pp. 581–585 (2003)
6. Guo, B., Wu, Y., Jiao, J., Lv, B., Zhou, F. Ma, Z., Sun, J.: Building entry loss model for 24 to 31 GHz band. In: International Symposium on Antennas and Propagation (ISAP) (2016)
7. Inomata, M., Yamada, W., Sasaki, M., Onizawa, T.: Outdoor-to-indoor path loss model for 8 to 37 GHz band. In: IEEE International Symposium on Antennas and Propagation (ISAP) (2015)
8. Inomata, M., Sasaki, M., Onizawa, T., Kitao, K., Imai, T.: Effect of reflected waves from outdoor buildings on outdoor-to-indoor path loss in 0.8 to 37 GHz band. In: IEEE International Symposium on Antennas and Propagation (ISAP) (2016)
9. Miura, Y., Oda, Y., Taga, T.: Outdoor-to-indoor propagation modelling with the identification of path passing through wall openings. In: IEEE International Symposium on Personal, Indoor and Mobile Radio Communications, vol. 1 (2002)
10. ITU-R: Propagation effects relating to terrestrial land mobile and broadcasting services in the VHF and UHF bands, ITU-R Recommendation P.1406-2 (2015)
11. ITU-R: Compilation of measurement data relating to building entry loss, ITU-R Recommendation P.2346-2 (2017)
12. ITU-R: Effects of building materials and structures on radiowave propagation above about 100 MHz, ITU-R Recommendation P.2040-1 (2015)

Upgrade of Electronic Security Fence System by Reduction of Vibration Noise by Wind Effect

Hiesik Kim[1(✉)] and Odgerel Ayurzana[2]

[1] Department of Electrical and Computer Engineer, University of Seoul, Seoul, Korea
drhskim@yahoo.com

[2] Department of Electronics, Mongolian University of Science and Technology, Ulaanbaatar, Mongolia
odgerel55@gmail.com

Abstract. Vibrational noise signal due to wind force on the electronic fence security system need to be overcome. Sensor cables were already fast fixed on to existing security fence perimeter. The wind noises generated by the strong wind were much problem by the real operation of electronical fence system. Wind noise signals receiving from sensor cables were processed and analyzed to improve the existing program in the DSP (TMS320F2812) microcontroller. The system collects signal amplitude, duration time and frequency spectrum to distinguish the wind effect from intrusion signal. The result gives the real or false alarm. Frequency analysis was done for each duration of N = 128 samples of input signal was using special algorithm of the fast Fourier transform through DSP microcontroller. The security system has been tested at border lines of Mongolia. A detection rate after apply the developed algorithm for reducing false alarms was improved up to 94%– 95% of reliable accuracy for real field application.

Keywords: Perimeter sensor · Electrostatic charge · Friction electricity Security fence · Invasion detection · Intrusion detection · False alarm

1 Introduction

1.1 Upgraded Electrical Security Fence

The digital electrical security systems with intrusion detection sensors become more critical problem for any security organization and countries. Many strategically important factories and facilities have been built recently by using rapid industrial it technology development. Security systems were developed with the advanced technologies. The electric charge sensor fence was developed as the prospected system in this field. It is the first passive sensor in the world. It uses the special algorithm of electric charge displacement that is based on friction electromotive force change of the passive sensor cable. The electric charge technique is based on the triboelectric effect. If we rub two different types of materials with each other, the electrostatic charges are

S. O. Hwang et al. (Eds.): ICGHIT 2018, LNEE 502, pp. 57–63, 2019.
https://doi.org/10.1007/978-981-13-0311-1_11

generated that are called the triboelectric effect. The physical protection security system can be realized by triboelectric effect. A simple telecommunication cable is used in the protection systems instead of the sensor transducer that is fastened to the various types of perimeter fences. Electrostatic charges are generated, when external force is applied to the sensor cable.

The ASM (Analog Sensing Module) [1, 2] detects the generated minimal charges and rings an alarm. The ASM doesn't generate any electric signals and electromagnetic fields. It only reacts against the change on electrostatic charge in specific range and is not affected by any external factors. Therefore there are no false alarms and perfect probability detection. Also the security detector does not generate the false alarm signals on and after exposure to the outdoor environment factors including humidity, rain, wind, fog, dust etc. the sensor cable reacts against the forced impact weighing 8–20 kg on the fence. It does not detect little forces including small animals hit the fence. The system sensitivity can be regulated by switches. Electric charge sensor fences are not harmful to the human body. These types of security systems have much more advantages than any other systems. It has simple installation and easy maintenance. And it was highly useful and more economical than any other electrical fence.

2 System Design and Its Solution

2.1 System Structure

Electrostatic charges are generated between the cable conductors and isolator when external force and impact are created by the intruders. Simple shielded 15 pair of telecommunication cable is used instead of the sensor transducer for sensing the external force and impact. Sensor cable is not connected to the power source. The lengths of sensor cable are limited up to 1000 m. Figure 1 shows the main operation diagram of the system.

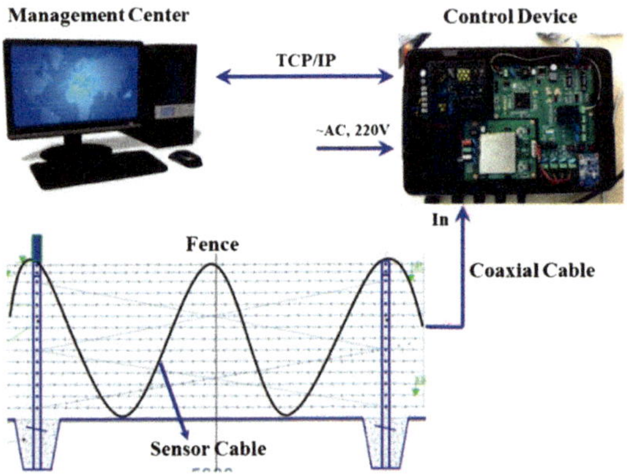

Fig. 1. Operation diagram of the electronic fence system

Coaxial cable is used for transferring generated minimal electrostatic charge between the sensor cable and control device. Because of the very high input impedance of the charge amplifier, the sensor must be connected to the amplifier input with low-noise cable [3]. This cable is specially treated to minimize triboelectric noise which is generated within the cable due to physical movement of the cable. The coaxial cable is necessary to affect an electrostatic shield around the high impedance input lead, precluding extraneous noise pickup.

Installation configuration of the sensor cable is dependent on the fence type, structure, and size, height, and installation weather conditions. System sensitivities can be adjusted by hardware method in the control device. Also sensitivities can be adjusted by software in the monitoring application program at the data center. Sensitivity is adjusted by less when sensor cable is fastened on flexible and moving fences. Conversely, sensitivity is adjusted by high when cable is fastened on rigid and less moving fences. The control device controls up to two security zones. Each zone are covered with 500 m sensor cable.

2.2 Operation Principle of a Control Device

The control device contains two main parts named as ASM (Analog Sensing Module) and SCM (Sensitivity Control Module). The ASM [1] detects the electrostatic charges on the passive sensor cable by intruder and raises the alarm. The ASM contains a charge sensitive device, voltage amplifier, signal shaping, filtering, and comparator. The charge sensitive device consists of the charge preamplifier and filters.

The SCM is the digital part of the control device that processes and analyses analog signals from a ASM using by TMS320F2812 32 bit DSP (Digital Signal Processing) microcontroller.

The input signal of the sensor cable is fluctuated due to outside environment effects (strong wind, storm) then the SCM analyses all conditions and adjusts sensitivity automatically. For instance SCM reduces sensitivity during strong wind and increases sensitivity during low wind. Also SCM analyses the input signal amplitude, duration time and frequency. These parameters are changed due to the wind effect. After analyzing input signal, system decides which alarms are real or false. Each N = 128 samples of input signal were analyzed special algorithm of the discrete Fourier transform by using Eq. (1).

$$X(k) = \sum_{n=0}^{N-1} x(n) * e^{\frac{-j2\pi k}{N}n}; \; 0 \le k \le N - 1 \tag{1}$$

x(n): Input signal

N = Count of sample (128)

The 1 Hz ± 0.5 Hz frequency is generated in the control device when someone forces by 8–20 kg impact on the fence. This is an invasion frequency. The band bass filter is designed in the ASM of control device [8, 9]. Other frequencies are generated due to strong wind effects (Fig. 2).

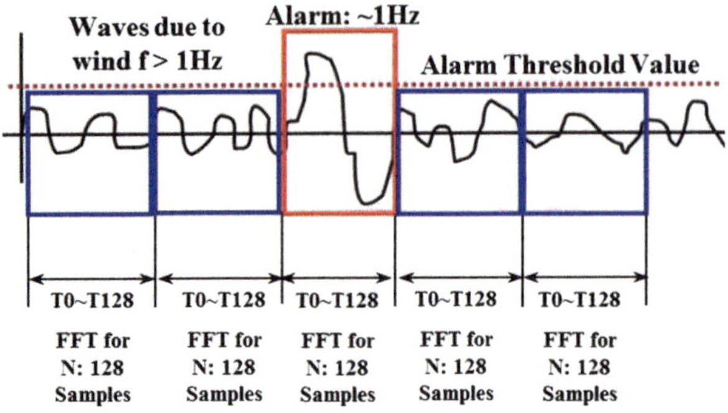

Fig. 2. Condition of fast Fourier transform in input signal

The alarm condition of the system is shown. Input analog signal of the sensor cable is sampled by 10 ms time step in each N = 128 samples using by ADC of the control device. Frequencies are defined every T = 1 ms * 128 = 1.28 s with the fast Fourier transform.

Real wind configuration is shown in Fig. 3. False alarms are generated when signal amplitude of strong wind exceeds a reference value.

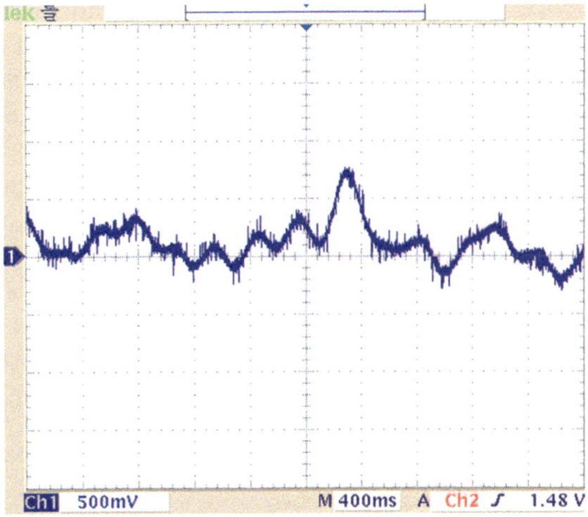

Fig. 3. Wave form of noise signal of wind force vibration on the electrical fence

Algorithm for reducing false alarms was developed. This algorithm reduces false alarms due to strong wind effect. ADC of control device processes N = 128 sample data by 10 ms steps. An algorithm reviews and analyses frequency value using FFT, maximum and minimum value of amplitudes.

- Vmax: Maximum value of amplitude in each duration time of 1.28 ms
- Vmin: Minimum value of amplitude in each duration time of 1.28 ms

- Fmax: Maximum frequency in each duration time of 1.28 ms by FFT
- Vpw: Value of plus wind level that is regulated by DIP switch on control device
- Vmw: Value of minus wind level that is regulated by DIP switch on control device
- Vpth: Plus threshold value
- Vmth: Minus threshold value

Then all these signal characteristics values are compared to the reference values. After that system decides which alarms are real or false. A detection rate was 85–88% before implementing algorithm for reducing false alarms. Detection rate of the system is improved up to 94–95% after implementing that algorithm. That means system does not alarm during strong wind. All data of the wind effect is saved to data center. We have to improve this algorithm using by saved data.

3 Experiment and Results

3.1 Experiment Field and Conditions

The perimeter security system has been tested at the border of Mongolia (2014.11.09–to present). The area has special weather conditions. In the north of Mongolia it is very cold and snow stormy during winter seasons. Also there is strong wind during spring and fall seasons.

Figure 4 shows installation of sensor cables in the selected special zones of experiment area. Sensor cable is fastened to the border fence by sine configuration. Sensor cable configuration is depended on fence types.

Fig. 4. Installation of sensor cable in the real field test

Two separated zones for testing security system were set up on the barbed fence. This type of fence is flexible and easy moving that is very sensitive in wind effect.

Experiment has been done two-three times in a day. Operators have pull and push barbed fence then alarms are generated on the data center. Monitoring application program registers all generated alarms and system conditions.

Table 1 shows some registered alarms in the data center from 2014.11.10 to 2015.04.25. As shown in experiment results, there are no false alarms when there are windless days. Alarms are generated when a dog and a cow touches the fence. Also some alarms are generated when bevy of pies and crows seat and fly on the barbed wire of the fence. This type of alarms are normal operation of system. But some false alarms are generated during snowstorm in 2015.01.05,06.

Table 1. List of all registered invasion alarms.

Alarm time	Alarm zone	Alarm reason	Operator
11/15/2014 12:3	1	Test	P. Davaadorj
11/28/2014 12:3	2	Test	S. Chuluun
12/16/2014 8:29	2	A dog touch	G. Olzii
12/27/2014 9:03	1	Test	P. Davaadorj
1/5/2015 2:06	2	Snowstorm	B. Shinebayar
1/5/2015 12:19	1	Snowstorm	B. Shinebayar
1/6/2015 14:12	2	Snowstorm	G. Olzii
1/8/2015 11:58	1	A cow touch	B. Shinebayar
1/13/2015 14:05	1	A bevy of pies	E. Erdene
1/16/2015 11:29	2	A bevy of crows	S. Chuluun
1/18/2015 15:13	1	A cow touch	B. Shinebayar
2/8/2015 11:58	2	Test	B. Shinebayar
2/23/2015 14:05	1	A dog touch	G. Olzii
3/11/2015 11:29	2	Test	P. Davaadorj
3/28/2015 15:13	1	A cow touch	B. Shinebayar
4/17/2015 13:35	1	Test	S. Chuluun
4/25/2015 14:05	2	Test	G. Olzii

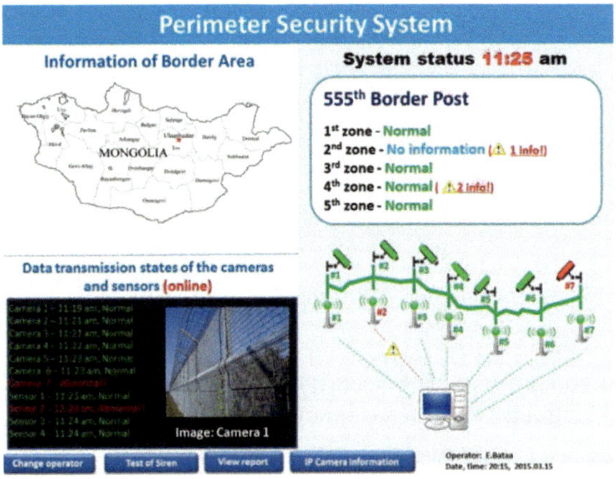

Fig. 5. Screen capture of fence security program

3.2 Monitoring Program of Security Fence

All log data is stored to the data center. Monitoring program receives all zones information by real time from control device and displyas these states. For example there are intruders' alarms, sensor and coaxial cable is cut or short, control device's cover open. A report can be printed and viewed with many options (Fig. 5).

4 Conclusions

The first version of the electrical fence security system was upgraded and applied in real field experiment under the hard and severe climate conditions of Mongolia. False alarms were the main problem by real operation of the fence security system. These false noise was generated due to strong wind force. In order to reduce false alarms, special algorithm was researched and implemented in the system. As shown in experiment results, alarms are generated when a dog get in under barbed fence and a cow touches the fence. Also some alarms are generated when bevy of pies and crows seat and fly on the barbed wire of the fence. These types of alarms are normal operation. But some false alarms are generated when snow is stormed continuously.

The precision detection rate of the invasion was improved up to 95% as seen from experimental data of the previous 5 months. A new algorithm for reducing wind noise problems was developed to increase positive detection rate. Two CCTV cameras of IP protocol was installed on to this security system to improve further comfortability.

When it is very cold weather, the sensor cable is frozen by outside temperature of −30 °C coldness. In this severe case, the electronic fence sensitivity was decreased. So it needs to be adjusted automatically to increase sensitivity range by 1–2 steps in very cold winter season.

Acknowledgment. This research was financially supported by the Project No. 20150126 2003 "Sensitive and Errorless Electrical Security Fence by Using Minimal Friction Electricity in Coaxial Cable" of the National Research Foundation of Korea and by the Project No. COR_01/2015 "High Sensitive and Errorless Security Fence System" of the Mongolian Foundation for Science and Technology.

References

1. Odgerel, A., Chol, L.Y.: Security system design based on the triboelectric effect. In: The 7th Conference on National Defense Technology, Korea University, Seoul, Korea, 07–08 July 2011
2. Kim, H., Yun, S.J., Ayurzana, O.: Implementation of intrusion monitoring system to operate optimim by using electronic security fence of friction electricity sensor. In: International Conference on ICS 2015, Gannin City, South Korea, 23–24 April 2015
3. Ayurzana, O., Kim, H.: Minimal electric charge detection device for fence security systems. In: MUSTAK 2015, Ulaanbaatar, Mongolia, 20–21 August 2015

Study of Semi Deterministic Model for Fifth-Generation (5G) Wireless Networks

Supachai Phaiboon[1(✉)] and Pisit Phokharatkul[2]

[1] Department of Electrical Engineering, Faculty of Engineering, Mahidol University, Buddhamonthon 4 Road, Buddhamonthon, Nakornpathom 73170, Thailand
supachai.pai@mahidol.ac.th
[2] Department of Electronic and Telecommunication Engineering, Faculty of Engineering, Kasem Bundit University, Pattanakarn Road, Suanluang, Bangkok 10250, Thailand

Abstract. This paper presents study of path loss models of fifth generation (5G) wireless communication systems. Propagation parameters such as path loss at reference distance (PL(d0)), path loss exponent (PLE) and standard deviation of the zero-mean Gaussian random both line-of-sight (LOS) and non line-of-sight (NLOS) are compared at the frequencies of 28, 38 and 73 GHz. Omni directional propagation large-scale path loss measured data from two downtown Cities are used. This paper also compares with semi deterministic models such as WI model and Xia model for present 4G network in order to develop the semi deterministic model for 5G networks.

Keywords: Empirical model · Semi deterministic model
Two downtown cities · Millimeter-wave Omni-directional path loss

1 Introduction

Mobile communication is moving to fifth generation (5G) at millimeter wave (mmWave) frequencies in order to provide multi-Gigabit-per second (Gbps) data rates to a mobile device for video and the Internet-of-Things (IoT). In order to install communication station, the propagation path loss models are one of important things. There are three types of the path loss models namely, empirical model, semi-deterministic model and deterministic model. The empirical models for 5G [1, 2] are widely used since they need only frequency, PLE and distance to compute the path loss for macro and micro cell planning while the deterministic models need the details of digitize map and material of buildings. Additionally they use a lot of time for computing. Finally the semi-deterministic models [3–9] are also widely used since they are not only need the parameters of the empirical model but also require some information about buildings such as dimension and type of them together with wide and direction of road. This model provides more accuracy path losses and is used for planning and solving the communication system. For semi-deterministic models, there are previous studies as follows; Xia et al. [3, 4] proposed path loss formulas for micro-cells in low-rise and high-rise building environments. Additionally, COST 231 WI model [5–7] is

S. O. Hwang et al. (Eds.): ICGHIT 2018, LNEE 502, pp. 64–69, 2019.
https://doi.org/10.1007/978-981-13-0311-1_12

also a popular prediction tool for micro cell environments. However, they need environment data base details. In case of the Xia model for non light of sight (NLOS), it requires a set of model parameters such as distance from the last roof top to receiver, average building height, and antenna heights as in our research [8, 9], while the WI model requires the road parameters such as height of buildings, width of the roads, building separation and road orientation with respect to the direct radio path. The aim of this research is a study the millimeter wave empirical path loss model also compares with semi deterministic models such as WI model and Xia model for present 4G network in order to develop the semi deterministic model for 5G networks in the next step.

2 Path Loss Models

2.1 A. Empirical Path Loss Model

This model does not utilize the parameter of specific area except the frequency and distance.

$$PL_{LN}(dB) = PL(d_0) + 10nlog_{10}\left(\frac{d}{d_0}\right) + x_\sigma \tag{1}$$

Where n is the path loss exponent, that indicates the attenuation rate at the distance (n = 2 for free space), $PL(d_0)$ is the path loss at a reference distance 1 m. and X_σ is a zero mean random variable, that have Gaussian distribution with standard deviation σ. The $PL(d_0)$ is found from Eq. (2)

$$PL_{FS}(dB) = 32.44 + 20log_{10}(f) + 20log_{10}(d) \tag{2}$$

It can find frequency f in unit of GHz and the distance d in meters. This case is valid for the free space and far-field region. If the ground reflection over an earth plain was included, the model becomes.

$$PL_{GR}(dB) = PL_{FS}(dB) - 20log_{10}\left[2\left|\sin\left(\frac{\beta h_T h_R}{d}\right)\right|\right] \tag{3}$$

h_T and h_R are the transmitter and receiver antenna heights, respectively. $\beta = 2\pi/\lambda$ and λ is the wavelength. This case is valid for large distances from the transmitter, and $\gg \beta h_T h_R$, that only depends on d, h_T *and* h_R, and the attenuation which correspond to the free space.

2.2 Semi-Deterministic Path Loss Model

There are two approach models which are used for a 2D map, are compared as follows:

– *Xia model*

The original Xia model for low-rise environments with one five story building was applied to predict path loss because it needed 2D maps of buildings for calculation. There are three routes for prediction namely, staircase route, transverse route and lateral route as shown in Fig. 1. A transmitter (Tx) was located on the street in the middle of a building block. The original Xia path loss formulas for all non line of sight cases were written as,

$$\begin{aligned} PL(d) = {} & [139.01 + 42.59 \log fG] - [14.97 + 4.99 \log fG]\operatorname{sgn}(\Delta h)\log(1 + |\Delta h|) \\ & + [40.67 - 4.57\operatorname{sgn}(\Delta h)\log(1 + |\Delta h|)]\log d \\ & + 20\log(\Delta hm/7.8) + 10\log(20/dh) \end{aligned} \quad (4)$$

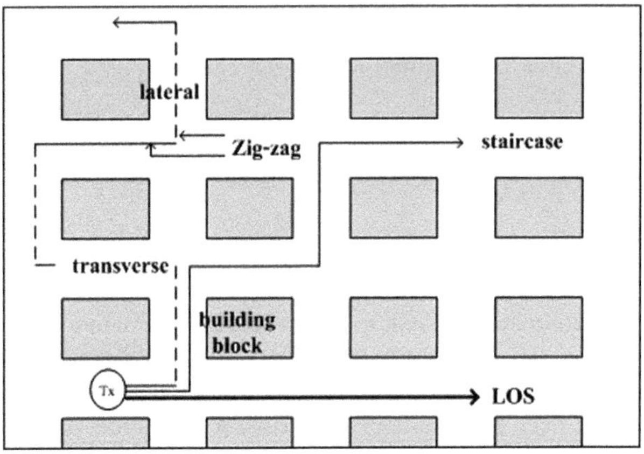

Fig. 1. Geometry of Xia model

Where d is the mobile distance from transmitter (km). [0.05 < d < 3], fG is the frequency (GHz). [0.9 < fG < 2], Δh is the relative height of transmitter to average building height (m). [–8 < Δh < 6], Δhm is the height of the last building relative to the mobile (m), dh is the distance of mobile from the last rooftop (m), hb is the transmitting antenna height from ground level (m), hm is the mobile antenna height from ground level (m) and λ is the wavelength (m).

– *WI model*

WI model is written in case of non line of sight as follows:

$$L(NLOS) = 32.4 + 20\log(f) + 20\log(d) + L(diff) + L(mult) + 3 \quad (5)$$

Where

$$L(diff) = -6.9 + 10\log(w) + 10\log(f) + 20\log(dhm) + Lori \quad (6)$$

When f is frequency (MHz), d is distance (km.), W is width of road (m.), hb is base station antenna height (m.), hm is mobile height (1.8 m.) hroof is average height of roof top, Δhm is distance between hm and hroof.

$$\begin{aligned} \text{Lori} &= -10 + 0.354\phi \text{ for } 0 \leq \phi < 35 \\ &= 2.5 + 0.075(\phi - 35) \text{ for } 35 \leq \phi < 55 \\ &= 4 - 0.114(\phi - 55) \text{ for } 55 \leq \phi < 90 \end{aligned} \tag{7}$$

where ϕ is the angle between incidences coming from base station and road

$$\text{L(mult)} = \text{k0} + \text{ka} + \text{kd}\log(\text{d}) + \text{kf}\log(\text{f}) - 9\log(\text{W}) \tag{8}$$

Where k0 = 0, kd = 18–15(Δhb/hroof), ka = 54–0.8 (Δhb), and kf = –4 + 0.7 [(f/925)–1] in case of sub-urban.

3 Measurement and Location

This paper used measured data in the dense urban environment around New York University's (NYU) Manhattan campus at both 28 GHz and 73 GHz [1, 2],and around the campus of The University of Texas at Austin (UTA) at 38 GHz. These measurements will be Omni directional large-scale line-of-sight (LOS) and non-line-of-sight (NLOS) directional measurements. In the NYU, the building density is about 65% while the building height is about 5–50 m. In the UTA, the building density is about 55% while the building height is about 5–30 m. The measurement procedure and description of the equipment can be found for details in [1]. The transmitting antenna heights of 28 GHz and 73 GHz are 7 and 17 m while the transmitting antenna heights of 38 GHz are 8 m, 23 m, and 36 m. The receiving antenna heights of 28 GHz and 38 GHz are 1.5 m while the receiving antenna heights of 73 GHz are 2.0 m and 4.06 m.

4 Results and Analysis

The empirical path loss model at frequencies of 28 GHz, 38 GHz and 73 GHz are respectively as following.

4.1 Empirical LOS

$$\begin{aligned} PL_{28GHz}(dB) &= 61.3 + 21log_{10}(d) \\ PL_{38GHz}(dB) &= 60.0 + 19log_{10}(d) \\ PL_{73GHz}(dB) &= 69.7 + 20log_{10}(d) \end{aligned}$$

4.2 Empirical NLOS

$$
\begin{aligned}
PL_{28GHz}(dB) &= 61.3 + 34log_{10}(d) \\
PL_{38GHz}(dB) &= 60.0 + 25log_{10}(d) \\
PL_{73GHz}(dB) &= 69.7 + 34log_{10}(d)
\end{aligned}
$$

4.3 Semi Deterministic Models

$$
\begin{aligned}
PL_{28GHz,NLOS,Xia}(dB) &= 53.5 + 51.8log_{10}(d) \\
PL_{38GHz,LOS,WI}(dB) &= 56.2 + 26log_{10}(d) \\
PL_{73GHz,NLOS,Xia}(dB) &= 61.9 + 58.2log_{10}(d)
\end{aligned}
$$

Note that the models are still applied for frequency range of 0.8 to 3 GHz. This needs correction factors to adjust these models. The Xia model provides over estimate path loss about 15% comparing with the measurement while the comparisons in this paper are only LOS situation for WI model. We also found an agreement for LOS while the NLOS of semi deterministic models have to adjust in order to meet good agreement for the fifth generation (5G) at millimeter wave (mmWave) frequencies.

In case of Xia model for NLOS, it requires a set of model parameters such as distance from the last roof top to receiver, average building height, and antenna heights. While the WI model require height of buildings, width of the roads, building separation and road orientation. These intonations are available in 3D map.

5 Conclusion

We present path loss models of fifth generation (5G) wireless communication systems. Propagation parameters such as path loss at reference distance (PL(d0)), path loss exponent (PLE) are calculated both line-of-sight (LOS) and non line-of-sight (NLOS) at the frequencies of 28, 38 and 73 GHz. We used measured data in the dense urban environment around New York University's (NYU) Manhattan campus at both 28 GHz and 73 GHz and around the campus of The University of Texas at Austin (UTA) at 38 GHz. This paper also compares with semi deterministic Xia and WI model. The results shown that the NLOS of semi deterministic models provide over estimate path loss from the measured data. So they have to adjust in order to meet good agreement for the fifth generation (5G) at millimeter wave (mmWave) frequencies.

References

1. Maccartney, G.R., et al.: mmWave omnidirectional path loss data for small cell 5G channel modeling. IEEE Access **3**, 1573–1580 (2015)
2. Rappaport, T.S., et al.: Overview of mmWave communication for 5G wireless networks. IEEE Trans. Antennas Propag. **65**(12), 6213–6230 (2017)
3. Xia, H.H.: A simplified model for prediction path loss in urban and suburban environments. IEEE Trans. Veh. Technol. **46**(4), 1040–1046 (1997)
4. Har, D., Xia, H.H., Bertoni, H.L.: Path-Loss prediction model for microcells. IEEE Trans. Veh. Technol. **48**(5), 1453–1462 (1999)
5. Walfisch, J., Bertoni, H.L.: A theoretical model of UHF propagation in urban environments. IEEE Trans. Ant. Prop. **36**(12), 1788–1796 (1988)
6. Ikegami, F., Yoshida, S., Takeuchi, T., Umehira, M.: Propagation factors controlling mean field strength on urban streets. IEEE Trans. Antennas Propag. **32**, 822–829 (1984)
7. Bhuvaneshwari, A., Hemalatha, R., Satyasavithri, T.: Semi deterministic hybrid model for path loss prediction improvement. Procedia Comput. Sci. **92**, 336–344 (2016)
8. Phaiboon, S., Phokharatkul, P.: Path loss prediction for low-rise buildings with image classification on 2-D aerial photographs. In: Progress in Electromagnetic Research-pier, vol. 95, pp. 135–152 (2009)
9. Phaiboon, S., Phokharatkul, P.: Comparison between mixing and pure Walfisch-Ikegami path loss models for cellular mobile communication network. In: Proceedings of PIERS, 23–27 Mar 2009, pp. 99–104 (2009)

Development of Integrated Monitoring System Based on IoT for Odor Reduction

Hye-Young Kim(✉)

School of Games, Major of Game Software, Hongik University, Jochiwon-eup, Chungcheongnam-do 339-701, Korea
hykim@hongik.ac.kr

Abstract. In this study, we developed an integrated monitoring system that can control the odor environment and reduce the odor by using the data from complex sensor devices for reducing odor in livestock facilities in IoT environment.

Keywords: IoT · Integrated monitoring system · Odor reduction

1 Introduction

Analysis of odor the evaluation of odor intensity is based on the method of indicating odor intensity, unpleasantness, and odor frequency, but most of them use human olfactory sense or collect air from odorous area. This analysis is time-consuming and difficult to find because it is difficult to analyze immediately. Therefore, in this study, we developed an integrated monitoring system that can control the odor environment and reduce the odor by using the data from the complex sensor to reduce odor in the livestock facility in the IoT environment.

2 Monitoring System

2.1 System Configuration

We made the MySQL database development environment on the Windows 10 operating system. The server model was developed using IOCP (Input/Output Completion Port) model. Our monitoring system stores the data of each enclosure received from the repeater in the MySql database through query and transmits the data.

We show our monitoring system configuration and UML of our monitoring system in Figs. 1 and 2, respectively.

S. O. Hwang et al. (Eds.): ICGHIT 2018, LNEE 502, pp. 70–72, 2019.
https://doi.org/10.1007/978-981-13-0311-1_13

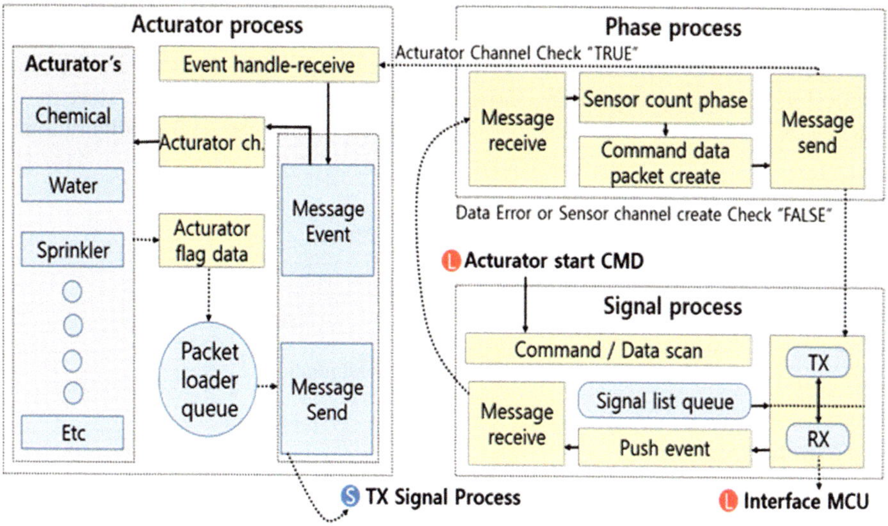

Fig. 1. System configuration

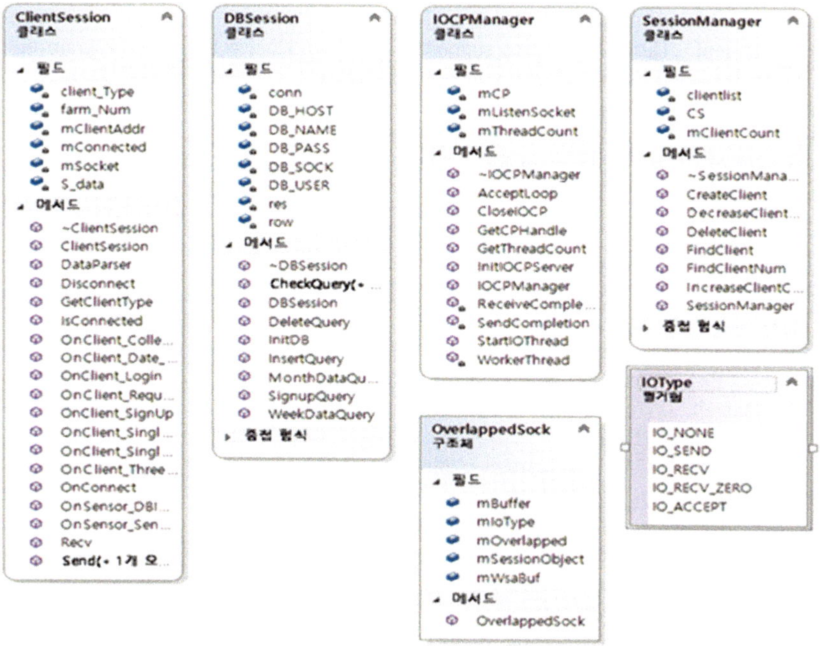

Fig. 2. UML of our system

2.2 Design of the Database System

We refer to FARM_CODE of FARM USER table in FARM DATE and ODORU DATA table separately, store ODORU DATA by date and store it on FARM DATE. In the Fig. 3, we show our designed database system configuration.

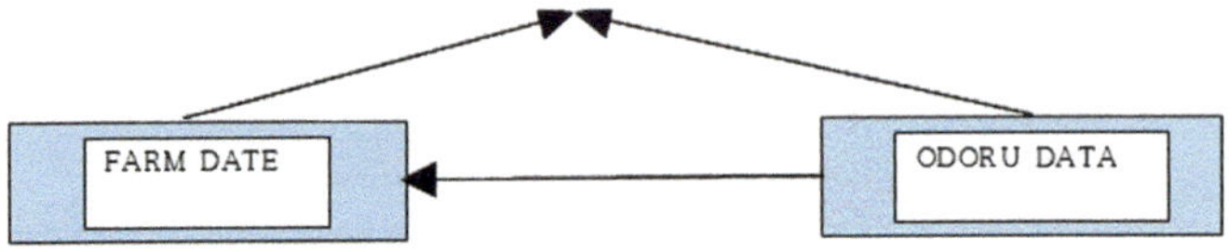

Fig. 3. Database system configuration

3 Conclusion

In this study, we developed a real-time monitoring program and database of camera image and odor measurement data such as livestock facilities, and developed a sensor node control and program for each housing, and developed an alarm system And integrated monitoring system was developed.

Acknowledgment. This research was supported by Basic Science Research Program through the National Research Foundation of Korea (NRF) funded by the Ministry of Science, ICT & Future Planning (No. 2016RIA2B4012386) also, this work (Grants No. 201602850002) was supported by Business for Cooperative R&D between Industry, Academy, and Research Institute funded Korea Small and Medium Business Administration in 2016 and this work was supported by 2018 Hongik University Research Fund.

References

1. Eason, G., Noble, B., Sneddon, I.N.: On certain integrals of Lipschitz-Hankel type involving products of Bessel functions. Phil. Trans. Roy. Soc. Lond. **A247**, 529–551 (1955). References
2. Clerk Maxwell, J.: A Treatise on Electricity and Magnetism, 3rd edn., vol. 2, pp. 68–73. Clarendon, Oxford (1892)
3. Jacobs, I.S., Bean, C.P.: Fine particles, thin films and exchange anisotropy. In: Rado, G.T., Suhl, H. (eds.) Magnetism, vol. III, pp. 271–350. Academic, New York (1963)
4. Elissa, K.: Title of paper if known (unpublished)
5. Nicole, R.: Title of paper with only first word capitalized. J. Name Stand. Abbrev. (in press)
6. Yorozu, Y., Hirano, M., Oka, K., Tagawa, Y.: Electron spectroscopy studies on magneto-optical media and plastic substrate interface. IEEE Transl. J. Magn. Jpn. **2**, 740–741 (1987). Digests 9th Annual Conference on Magnetics Japan, p. 301 (1982)
7. Young, M.: The Technical Writer's Handbook. University Science, Mill Valley (1989)

Classifying News Articles Using Feature Similarity K Nearest Neighbor

Taeho Jo(✉)

Hongik University, Sejong 30016, South Korea
tjo018@hongik.ac.kr

Abstract. This research proposes the KNN (K Nearest Neighbor) which computes the similarity between data items considering features or attributes as well as one to one values. The assumption of the independency among attributes is the violation against the reality especially in the text classification where words are used as features of texts. In this research, we define the similarity measure which considers both attributes and attribute values, modify the traditional version of KNN using the similarity measure, and apply it to the task of text classification. As benefits from this research, it provides the more compact representations of texts and the better performance. Therefore, the goal of this research is to implement the text categorization system with its more efficient data representations and better performance.

Keywords: Feature value similarity · Feature similarity K Nearest Neighbor · Text categorization

1 Introduction

Text categorization refers to the process of assigning a topic or category to each text, as an instance of pattern classification. Its preliminary task is to predefine topics or categories and allocate texts to each of them as the sample data. By learning the sample labeled texts, the classification capacity which is given as symbolic rules, equations or/and statistical model parameters, is constructed. Subsequently, novice texts are classified based on the constructed classification capacity. Even if various types of text categorization are available, the scope of this research is restricted to only hard text categorization where each text is classified into only one topic or category, exclusively.

Some problems are caused by encoding texts into numerical vectors and computing their similarities based on attribute values. Many features are required for maintaining the system robustness in encoding texts into numerical vectors because each feature covers very small proportion of documents [3]. The dominance of zero values in each numerical vector becomes the poor environment for computing their similarities because of the very weak discrimination among numerical vectors [3]. The assumption that the features are independent of each other violates against the reality [1]. Hence, in this research, we consider both

S. O. Hwang et al. (Eds.): ICGHIT 2018, LNEE 502, pp. 73–78, 2019.
https://doi.org/10.1007/978-981-13-0311-1_14

features and feature values for compute the similarity between two numerical vectors as the challenge against the problems.

Let us mention what we propose in this research as the idea. In this research, we assume that words are used as features for representing texts into numerical vectors, and attempt to consider their semantic relations. Based on the semantic relations among words, we define the similarity measure which considers both the feature similarity and the feature value similarity between representations of texts. Using the similarity measure, we modify the KNN into the version which computes the similarity between items based on the both similarities, and apply it to the text categorization, as the approach. Hence, we obtain the more discrimination among numerical vectors even in the sparse distribution, and the reduced number of features as the benefits from this research.

Let us mention the benefits expected from this research, based on the above proposals. From this research, we expect the potential possibility of reducing the dimensions of numerical vectors representing texts, by considering the similarity among features as well as among feature values. We may reduce the information loss in computing the similarity between texts by reflecting the semantic similarity among words which are given as features. We may expect the improved tolerance to the problems from the sparse distributions of numerical vectors, because the similarity between attributes is considered. Therefore, we expect both the better performance of the text categorization and the more efficient representations of texts from this research.

2 Previous Works

Let us mention the popularity of encoding texts into numerical vectors, and the proposal and the application of string kernels as the solution to the above problems. In 2002, Sebastiani presented the numerical vectors are the standard representations of texts in applying the machine learning algorithms to the text classifications [4]. In 2002, Lodhi et al. proposed the string kernel as a kernel function of raw texts in using the SVM (Support Vector Machine) to the text classification [5]. In 2004, Lesile et al. used the version of SVM which proposed by Lodhi et al. to the protein classification [6]. In 2004, Kate and Mooney used also the SVM version for classifying sentences by their meanings [7].

It was proposed that texts are encoded into tables instead of numerical vectors, as the solutions to the above problems. In 2008, Jo and Cho proposed the table matching algorithm as the approach to text classification [8]. In 2008, Jo applied also his proposed approach to the text clustering, as well as the text categorization [11]. In 2011, Jo described as the technique of automatic text classification in his patent document [9]. In 2015, Jo improved the table matching algorithm into its more stable version [10].

Previously, it was proposed that texts should be encoded into string vectors as other structured forms. In 2008, Jo modified the k means algorithm into the version which processes string vectors as the approach to the text clustering [11]. In 2010, Jo modified the two supervised learning algorithms, the KNN and the

SVM, into the version as the improved approaches to the text classification [12]. In 2010, Jo proposed the unsupervised neural networks, called Neural Text Self Organizer, which receives the string vector as its input data [13]. In 2010, Jo applied the supervised neural networks, called Neural Text Categorizer, which gets a string vector as its input, as the approach to the text classification [14].

3 Proposed Approach

3.1 Text Encoding

Words are generated as feature candidates from the corpus through text indexing. Texts in the collection are concatenated into a big text and it is segmented into tokens by white spaces or punctuation marks. Each token is transformed into its root form through the stemming process; the verbs and nouns are transformed into their root forms and singular forms, respectively. For more efficiency, the stop words which functions only grammatically such as conjunctions, pronouns, and section, are removed from the stemmed tokens. A list of words which consists of verbs, nouns, and adjectives are extracted as the feature candidates.

Once some words are selected as attributes, we need to consider the schemes of defining a value to each attribute. To each attribute, we may assign a binary value indicating whether the word present in the text, or not. We may use the relative frequency of each word in the text. The weight of word to each attribute which is computed by Eq. (1) may be used as a feature value.

$$w_i = TF_i(\log_2 N - \log_2 DF_i + 1) \tag{1}$$

where TF_i is the total frequency in the given text, DF_i is the total number of documents including the word, and N is the total number of documents in the corpus. Therefore, the attributes values of a numerical vector which represent a text are relationships between the word and the texts which are selected as features.

Once some words are selected as attributes, we need to consider the schemes of defining a value to each attribute. To each attribute, we may assign a binary value indicating whether the word present in the text, or not. We may use the relative frequency of each word in the text. The weight of word to each attribute which is computed by Eq. (1) may be used as a feature value. Therefore, the attributes values of a numerical vector which represent a text are relationships between the word and the texts which are selected as features.

3.2 Feature Similarity

Words are given as features for encoding texts into numerical vectors. They are dependent on others rather than independent ones which are assumed in the traditional classifiers, especially in Naive Bayes [1]. Previously, various schemes of computing the semantic similarity between words were developed [2]. We need to assign nonzero similarity between two numerical vectors where non-zero elements are given to different features with their high similarity. It is expected to improve the discriminations among sparse vectors by considering the similarity among features (Fig. 1).

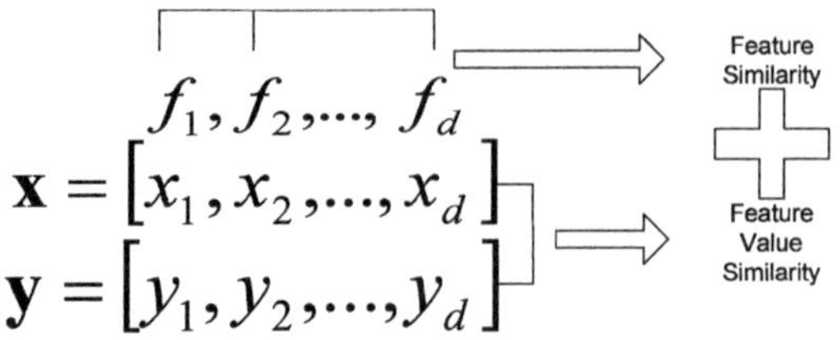

Fig. 1. The combination of feature and feature value similarity

We may build the similarity matrix among features automatically from a corpus. From the corpus, we extract easily a list of words. We compute the similarity between two texts by Eq. (2),

$$s_{ij} = sim(t_i, t_j) = \frac{2 \times df(t_i, t_j)}{df(t_i) + df(t_j)} \tag{2}$$

where $df(t_i, t_j)$ is the number of texts which include both words, t_i and t_j, and $df(t_i)$ is the number of texts which includes the word, t_i. We build the similarity matrix which is consists of similarities between text identifiers given as features as follows:

$$S = \begin{pmatrix} s_{11} & s_{12} & \dots & s_{1d} \\ s_{21} & s_{22} & \dots & s_{2d} \\ \vdots & \vdots & \ddots & \vdots \\ s_{d1} & s_{d2} & \dots & s_{dd} \end{pmatrix}$$

The rows and columns in the above matrix, S, correspond to the d words which are selected as the features.

The words, $t_1, t_2, ..., t_d$ are given as the features, and the two texts, d_1 and d_2 are encoded into the two numerical vectors as follows:

$$d_1 = [w_{11}, w_{12}, ..., w_{1d}]$$
$$d_2 = [w_{21}, w_{22}, ..., w_{2d}].$$

The features, $t_1, t_2, ..., t_d$ are defined through the process which was described in Sect. 3.1. We construct the d by d matrix as the similarity matrix of features by the process mentioned above. The similarity between the two vectors are computed with the assumption of availability of the feature similarities, by Eq. (3),

$$sim(d_1, d_2) = \frac{\sum_{i=1}^{d} \sum_{j=1}^{d} s_{ij} w_{1i} w_{2j}}{d \cdot \|d_1\| \cdot \|d_2\|} \tag{3}$$

where $\|d_1\| = \sqrt{\sum_{i=1}^{d} w_{1i}^2}$ and $\|d_2\| = \sqrt{\sum_{i=1}^{d} w_{2i}^2}$. We get the value of s_{ij} by Eq. (2).

3.3 Proposed Version of KNN

The traditional KNN version is illustrated in Fig. 2. The sample texts which are labeled with the positive class or the negative class are encoded into numerical vectors. The similarities of the numerical vector which represents a novice word with those representing sample words are computed using the Euclidean distance or the cosine similarity. The k most similar sample words are selected as the k nearest neighbors and the label of the novice entity is decided by voting their labels. However, note that the traditional KNN version is very fragile in computing the similarity between very sparse numerical vectors.

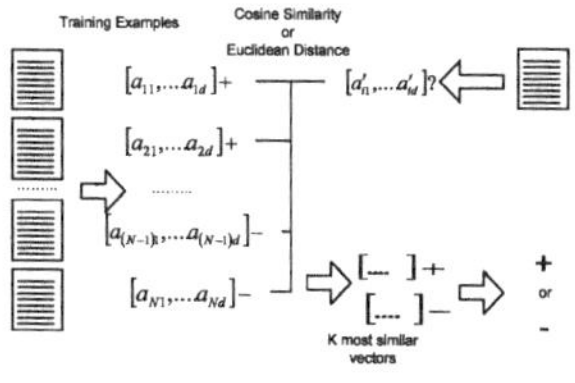

Fig. 2. The traditional version of KNN

The proposed KNN version is illustrated in Fig. 3. Like the traditional version, a text is given as an input and it is encoded into a numerical vector. The similarities of the novice text with the sample ones are computed by Eq. (3) which was presented in Sect. 3.2. Like the traditional version, k most similar samples are selected as the nearest neighbors, and the label of the novice is decided by voting their labels. The scheme of computing the similarity between numerical vectors is the essential difference between the two versions.

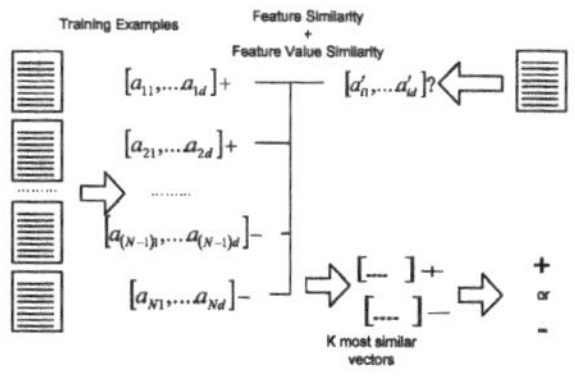

Fig. 3. The proposed version of KNN

4 Conclusion

Let us consider the remaining tasks for doing the further research. The proposed approach should be applied and validated in the specialized domains:

engineering, medicine, science, and law, and it should be customized to the suitable version. We may consider similarities among only some essential features rather than among all features, to cut down the computation time. We develop and combine various schemes of computing the similarities among features. By adopting the proposed approach, we will develop the text categorization system as a real version.

Acknowledgement. The research was supported by the International Science and Business Belt Program through the Ministry of Science and ICT (2017K000451).

References

1. Mitchell, T.: Machine Learning. McGraw-Hill, New York (1997)
2. Manning, C.D., Schutze, H.: Foundations of Statistical Natural Language Processing. MIT Press, Cambridge (1999)
3. Jo, T.: The implementation of dynamic document organization using text categorization and text clustering. Ph.D. Dissertation of University of Ottawa (2006)
4. Sebastiani, F.: Machine learning in automated text categorization. ACM Comput. Survey **34**, 1–47 (2002)
5. Lodhi, H., Saunders, C., Shawe-Taylor, J., Cristianini, N., Watkins, C.: Text classification with string kernels. J. Mach. Learn. Res. **2**, 419–444 (2002)
6. Leslie, C.S., Eskin, E., Cohen, A., Weston, J., Noble, W.S.: Mismatch string kernels for discriminative protein classification. Bioinformatics **20**, 467–476 (2004)
7. Kate, R.J., Mooney, R.J.: Using string kernels for learning semantic parsers. In: Proceedings of the 21st International Conference on Computational Linguistics and the 44th annual meeting of the Association for Computational Linguistics, pp. 913–920 (2006)
8. Jo, T., Cho, D.: Index based approach for text categorization. Int. J. Math. Comput. Simul. **2**, 127–132 (2008)
9. Jo, T.: Device and method for categorizing electronic document automatically. Patent Document, 10-2009-0041272, 10-1071495 (2011)
10. Jo, T.: Normalized table matching algorithm as approach to text categorization. Soft. Comput. **19**, 839–849 (2015)
11. Jo, T.: Inverted index based modified version of k-means algorithm for text clustering. J. Inf. Process. Syst. **4**, 67–76 (2008)
12. Jo, T.: Representationof texts into string vectors for text categorization. J. Comput. Sci. Eng. **4**, 110–127 (2010)
13. Jo, T.: NTSO (Neural Text Self Organizer): a new neural network for text clustering. J. Netw. Technol. **1**, 31–43 (2010)
14. Jo, T.: NTC (Neural Text Categorizer): neural network for text categorization. Int. J. Inf. Stud. **2**, 83–96 (2010)

A Knowledge Mining and Ontology Constructing Technology Oriented on Massive Social Security Policy Documents

Gang Liu(✉), Lian Sun, and Weiping Fu

Harbin Engineering University, Nantong St. 145, Harbin, China
liugang@hrbeu.edu.cn

Abstract. To address the problem brought from enormous policy documents and complex management in the social security domain, the article uses ontology as the way of representing and storing knowledge. The article constructs the framework of ontology through manual work so that it can ensure the relative accuracy of the ontology structure. Then it achieves the automatic ontology expansion based on the inclusion relationship of property sets or operational object sets. The article uses a semi-automatic method that extracts hierarchical concepts and non-hierarchical concepts from domain thesaurus by using the method combining statistics with rules to construct the ontology. Besides constructing the ontology, the article proposes the concepts of concept phrase vector model and high frequency characteristics phrase vector model. The experiment result indicates that ontology semi-automatic construction process can help experts to construct the social security ontology effectively oriented on collections of policy documents and is a considerable reference for the construction of ontology in other domains.

Keywords: Knowledge mining · Ontology expansion
Text fragment extraction · Semi-automatic construction

1 Introduction

In this paper, we introduce a novel method for Chinese ontology construction. It characterizes the domain text fragments by natural language processing based on its property set and operation object set.

There are two main types of the present method: constructing the ontology in artificial way, or by means of statistics. In this paper, we used a semi-automatic construction method of ontology (Ren et al. 2017).

2 Ontology Construction Preprocessing

Here we collect relevant data in the domain of Chinese social security by artificial way to build the required text collection. And these knowledge can mainly be divided into structured and unstructured and semi-structured knowledge (Du et al. 2006). The process can be described as follow.

S. O. Hwang et al. (Eds.): ICGHIT 2018, LNEE 502, pp. 79–85, 2019.
https://doi.org/10.1007/978-981-13-0311-1_15

Segmentation. Based on dictionary and grammars, the article segments the corpus collected through the algorithm invented by CAS (Chinese Academy of Science) (Liu et al. 2008).

Keyword Extraction Based on DDCF Evaluation Index. The method of keyword extraction can be learned from the method put forward by Chin-Ming Hong (Hong et al. 2009). DDCF (the difference of DCF) represents the diversity between the ith word composed of n words and The jth word composed of n + 1 words.

Fragment Extraction. Fragment is a keyword sequence gained through the above process. The article extracts fragments into triples consisted of a subject, a verb, and an object.

Initial Ontology Construction. Certain domain experts should establish a framework which contains most of the domain concepts, the concepts' operation object set, and the link between them. Knowledge Mining: Ontology Construction Based on Operated Set.

2.1 Ontology Construction Based on Property Set and Operational Object Set

Construction of Property Set and Operation Object Set. In the triples, a subject *C* stands for a concept. All verbs of *C* in different triples consist the property set of *C* (*PSetC*), while the objects consist the operational object set (*ObjSetC*).

The Update of The Property Set and The ObjSet. Refer to Fig. 1:

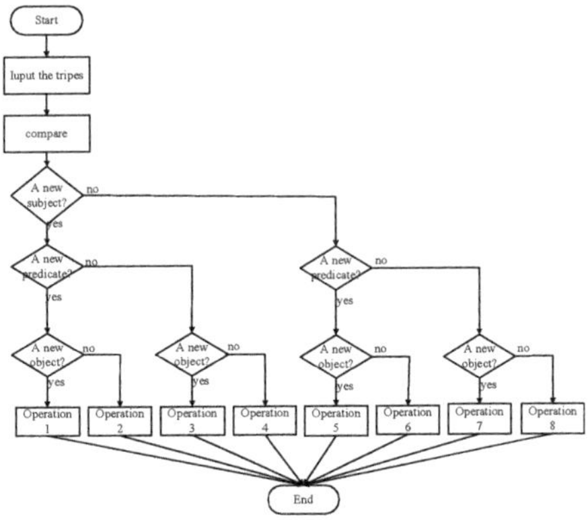

Fig. 1. The updating process

2.2 The Insertion of a New Node

The property similarity-computing algorithm mentioned by Yajuan Wu, etc. (Wu et al. 2009) take the factors of word similarity and *POS* into consideration, the calculation method is (1).

$$SimP(a_1, a_2) = \frac{1}{2}SimPN(a_1, a_2) + \frac{1}{2}SimPD(a_1, a_2) \tag{1}$$

a1 and are two concepts, $SimPN(a_1, a_2)$ stands for the word similarity of the property, $SimPD(a_1, a_2)$ represents the type similarity. (2) needs reconstruction for that the comparison objects of the article are the property nouns and the operational objects. Let $Sim(a_1, a_2)$ stand for the operational object word similarity, then the similarity between the property and the operational object could be described as (3).

$$Sim(a_1, a_2) = \frac{m}{n} \tag{2}$$

m is the same component numbers of the two vectors, n is the dimension of the longer between the two vectors.

Because the vector representation algorithm only considers the *ObjSetC* of the two concepts without using contextual information, a weight coefficient *W* should be added to (6), thus the calculating method is shown as (4):

$$W = \frac{Freq(a_1 \cap a_2)}{Freq(a_1) + Freq(a_2)} \tag{3}$$

$Freq(a_1 \cap a_2)$ indicates the simultaneous frequency of a_1 and a_2, $Freq(a_1)$ denotes the frequency of a_1. Similarity calculation could improve the accuracy with the help of *W*, the reconstructed similarity formula is shown as (5):

$$Sim(a_1, a_2) = \frac{a_1 \times a_2}{|a_1| \cdot |a_2|} \cdot \frac{Freq(a_1 \cap a_2)}{Freq\, a_1 + Freq\, a_2} \tag{4}$$

C_{jj} indicates the $j - th$ concept in the level I, $C_{jj}P$ is the parent of C_{jj}, $C_{jj}Child$ is the child of C_{jj}, R_1 represents the similarity threshold of the larger similarity result, R_2 represents the similarity threshold of the smaller similarity result.

3 Applications Based on Domain Ontology

3.1 Text Clustering Based on Domain Ontology

The article preprocesses the social security and gives an algorithm to optimize the high dimension feature vector space and practice the similarity of the texts by ontology.

Text Vectorization and Similarity Calculation. The article calculates words' weight by TF-IDF and selects the word with high weight as one of the words in the candidate phrase according to the threshold of weight.

The algorithm of generating the CVSM (Concept Vector Space Model) and HFVSM (High Frequency Vector Space Model) is as follow:

```
Algorithm 1: The Generation Of CVSM and HFVSM.
Input: Candidate Words, Word Frequency, Ontology, Threshold of Words'
Frequency
For each word in candidatWords
   concepts ← match(word , Onto);
   If(null != concepts) Then concepts.Add(concepts);  candidat-
Words.Remove(word);
   Else if(word.freq > maxFreq) Then highFreqWords.Add(word); candi-
datWords.Remove(word);
   End if
End for
Output: Concepts, High Frequency Words
```

4 Experiment Results and Analysis

4.1 Results of Ontology Constructing Experiments

Social Security Domain Corpus Collection. The article collected corpus from social security policies. The numbers of policies are 1210.

Keyword Extraction. Using TFIDF to extract keywords and using the evaluation index of DCF model and DDCF to assess the accuracy and recall rate (Tables 1 and 2).

Table 1. The result based on 720 texts

	TFIDF	DDCF
The accurate words	38960	42148
The whole words	65744	59230
Words given by experts	57453	57453
The accuracy rate	59.26%	71.16%
The recall rate	67.81%	73.36%

Fragment Extraction. Extract the most commonly used word combination according to the fragment extraction technology from the fragments after the keyword selection.

Table 2. The result based on 1210 texts

	TFIDF	DDCF
The accurate words	48706	62034
The whole words	103520	86278
Words given by experts	81667	81667
The accuracy rate	47.05%	71.90%
The recall rate	59.64%	75.96%

4.2 Results of Modular Experiments

In this section, the article at first circularly extends ontology based on domain thesaurus. This experiment circularly extends ontology four times in total (Table 3).

Table 3. Iteration Information

	Fundamental ontology	The first	The second	The third	The fourth
Concepts	20	50	118	289	363
Properties	3	5	8	10	11
Concepts not in ontology	1205	1295	1600	582	0
Residual words in corpus	5898	5630	4563	2844	1975

With the cycling times increasing Ontology concepts and attributes expands and the number of remaining words declines. With the expansion of ontology atomic concepts, relevant concepts increase. The number of relevant concepts decreases as the atomic concept and hierarchical concept constantly improve.

4.3 Experiments of Text Clustering

Compare of Methods Based on Ontology and VSM. In this section, the article uses more than 500 documents as test files. These documents can be classified as 6 types. The article evaluates the results of clustering by entropy and purity.
The entropy and purity comparison be found in Table 4.

Table 4. Comparison between VSM and ontology-based clustering method

	7 catalogs		13 catalogs		25 catalogs	
	Entropy	Purity	Entropy	Purity	Entropy	Purity
VSM	0.8723	0.4813	0.8532	0.5214	0.7763	0.5842
Ontology	0.6742	0.6237	0.5264	0.6776	0.6154	0.6845

It can be seen from Table 4 that ontology clustering method is better than VSM method. And the effect of clustering changes as the refinement of the cluster.

4.4 Results of Text Clustering Experiments

As we can see from Fig. 2, dimensions of concept phrase vector model and high frequency words vector model are smaller than the dimension of VSM model. The efficiency of clustering is far higher than the efficiency of traditional method.

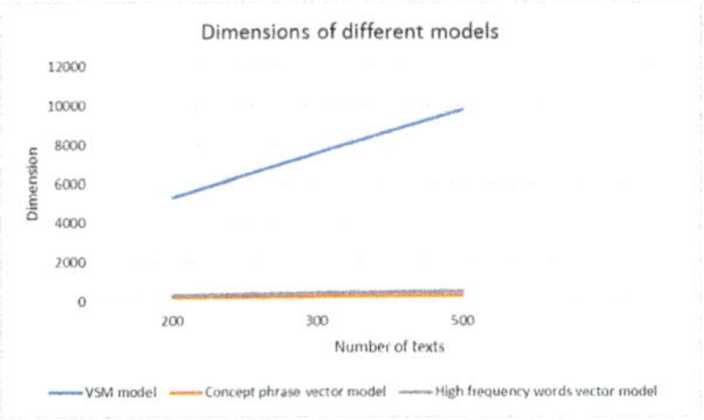

Fig. 2. The comparison of different models

5 Conclusion and Outlook

To solve the problem brought from the enormous social security policy documents and the complex management, the article constructs an ontology combined with the social security domain under the guidance of the seven-step method.

However, there are still some problems. Methods of constructing an ontology mentioned in this paper are not clear enough on operational details. Besides, this paper only discussed constructing methods based on unstructured texts, so the focus of further research is to process structured data and semi-structured data.

References

Pérez, A., Benjamins, V.: Overview of knowledge sharing and reuse components: ontologies and problem-solving methods. In: KRR5, Sweden, pp. 1–15 (1999)

Uschold, M., King, M.: Towards a methodology for building ontologies. In: IJCAI 95 Workshop on Basic Ontological Issues in Knowledge Sharing, pp. 1–13 (1995)

Jain, V.: Conversational ontology construction based on human emotions. In: CCTD, pp. 186–188 (2010)

Kong, J.: Ontology learning for Chinese information organization and knowledge discovery in ethnology and anthropology. Data Sci. J. **6**, 500–510 (2007)

Ren, F., Shen, J., Sun, B., Zhu, J.: A review for domain ontology construction from text. Chin. J. Comput. **40**, 3 (2017). (in Chinese)

Hong, C., Chen, C., Chiu, C.: Automatic extraction of new words based on Google News corpora for supporting lexicon-based Chinese word segmentation systems. Expert Syst. Appl. **36**(2), 3641–3651 (2009)

Shih, C.W.: Enhancement of domain ontology construction using a crystallizing approach. Expert Syst. Appl. **38**(6), 7544–7557 (2011)

Wu, Y., Chen, R., Shang, F.: A new ontology construction method based on similarity calculation. Comput. Appl. Res. **26**(003), 870–872 (2009). (in Chinese)

Sun, J., Huang, G., Luo, J.: An improved concept similarity algorithm. Comput. Eng. Appl. **45** (5), 154–156 (2009). (in Chinese)

Cai, W., Huang, S., Lv, T., Liu, L.: The social security level 3 linkage networked auditing mode research of practicing audit "immune system" function. Mod. Audit Account. **8**, 15 (2012). (in Chinese)

Sun, B., Liu, L.: A method of computing the semantic similarity of sentences based on HowNet. Comput. Digit. Eng. **42**(2), 187–209 (2014). (in Chinese)

Detection of GUI Elements on Sketch Images Using Object Detector Based on Deep Neural Networks

Young-Sun Yun[1(✉)], Jinman Jung[1], Seongbae Eun[1], Sun-Sup So[2], and Junyoung Heo[3]

[1] Hannam University, Daejeon, Republic of Korea
{ysyun,jmjung,sbeun}@hnu.kr
[2] Kongju National University, Cheoan, Republic of Korea
triples@kongju.ac.kr
[3] Hansung University, Seoul, Republic of Korea
jyheo@hansung.ac.kr

Abstract. Graphical user interface (GUI) is very important to interact with software users. In many studies, therefore, they are trying to convert GUI elements (or widgets) to code or to describe formally its structure by help of domain knowledge or machine learning based algorithms. In this paper, we adopted object detection based on deep neural networks that finds GUI elements by integration of localization and classification. After the successfully detection of GUI components, we will describe the objects as the hierarchical structure and transform those to appropriate codes by synthetic or machine learning algorithms.

Keywords: GUI elements detection · Object detector
Deep neural networks · Sketch image · Sketch to code

1 Introduction

GUI elements play an important role in most software applications. It allows software users to interact with the software functions and to perform the designated tasks in an intuitive and effective way. Designing its components and defining their functions are an important part of software application programming. There are many tools to support GUI programs, but, the program developers should learn their usages to compose GUI elements besides of program developing. Moreover, the programming languages to implement similar GUIs are different along to target platforms. The typical language-platform pairs are Swift/Objective C on iOS, Java on Android, and C/C++ with HTML5 on Tizen OS, etc. If the developer wants to transfer some application to other platforms, he/she repeatedly will code the UI and transforms the platform dependent codes. Therefore, the developers hope to convert or code some platform dependent modules automatically. In our research, the final goal is the development of

S. O. Hwang et al. (Eds.): ICGHIT 2018, LNEE 502, pp. 86–90, 2019.
https://doi.org/10.1007/978-981-13-0311-1_16

(semi-) automatic code conversion and building of GUI elements. As a starting point, we propose the GUI elements detection based on deep neural networks (DNNs) to describe the objects hierarchically in this paper. In the previous studies, most works relied on domain specific knowledge or generate the computer codes directly using combination of convolutional and recurrent neural networks (CNNs and RNNs). However, we suggest the object detector based recognition method of GUI elements.

2 Related Works

The detection of GUI elements is used for modeling objects using formal grammars [1] or data capturing the interactions between the software users and GUIs [2]. The recent study automatically generates platform dependent codes from a GUI screenshot using machine learning techniques [3].

Reference [1] used simple features, such as line, text, circle, etc. to recognize the graphic object and described its result as the form of a first order logic. These basic features are combined and augmented by a set of geometric relations and transformed to non-terminal symbols like Rectangle, TextCluster etc. They created object grammars for groupboxes, pushbuttons, radiobuttons, droplists, and checkboxes etc. and built the dependency graph of the grammar to describe the objects and to share them. Pham et al. proposed a Minion framework to address the core problems of mining visual log of software [2]. The framework contains object recognition and tracking algorithm to recognize GUI elements, to infer the changes occurring on those elements, and to generate the corresponding sequence of GUI interactions. For the recognition of GUI elements, they employed several algorithms based on computer vision and recognition rules manually designed for GUI elements with heuristics related to the elements.

On the contrary to the previous synthetic works, Beltramelli combined the image captioning technique and language modeling of textual description based on CNNs and RNNs, respectively [3]. He considered a generating computer code written from a GUI screenshot as a generating English text descriptions from a scene photography. In other words, he produced a variable-length strings of tokens from pixel values.

In this paper, we adopted the object detection approach using DNN rather than image captioning based CNN and RNN or synthetic knowledge based approaches. Object detection is the process that detects instances of real-world objects of a certain class in images or videos. Object detection algorithms typically use extracted features and learning algorithms to recognize instances of an object category. From [4], many studies apply CNN classification results on ImageNet to generalized object detection techniques. In the early study, object detection algorithm uses the separated process of localizations and annotations of detection data [4]. But, nowadays, many studies integrate the localization and detection approaches together within a single neural network. Therefore, they can achieve the real-time object detection with high accuracy.

3 Proposed System

In this section, we propose the automatic code generating system from a given sketch image with hierarchical object description model. The previous works only depend on semantic domain knowledge, or they pursue end-to-end algorithm that gets the screenshot image and generates the given platform codes. But, in the proposed system, we at first translate the GUI objects from sketch image to hierarchical object descriptions with object detection algorithm and next generate the GUI descriptions depended on the platforms through several managers. The overall system and its components are displayed in the Fig. 1.

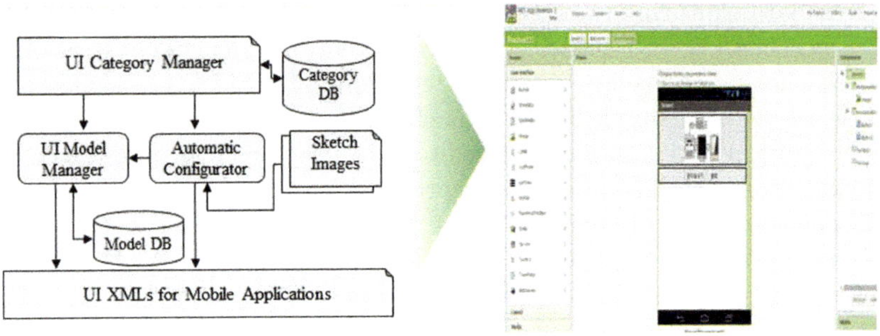

Fig. 1. Overall system and its component

3.1 Data Preparation

We designed 50 mimicked sketch images including about 600 GUI elements from the screenshot images gathered on Internet for preliminary study. To recognize the graphic objects, we first marked the GUI elements manually and save the bounding box and labels corresponding to the given images. The annotating tool is found at [6]. College students being familiar with GUI did annotate the components with bounding boxes and labels. LabelImg annotation tool is implemented with python and Qt. The resulting annotations are saved as XML file in PASCAL VOC format. The sample mimicked sketch image and the object annotation process are shown in Fig. 2.

3.2 Object Detection

To find the graphic objects in sketch images, we employed YOLO real time object detection technique. YOLO [5] applied a single neural network to the full image in contrary to the previous works which use multiple location and scale information. YOLO divides the image into regions and predicts bounding boxes and probabilities for each region. These bounding boxes are weighted by

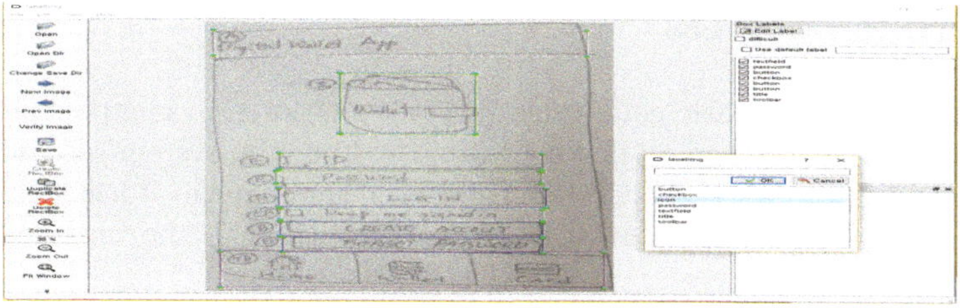

Fig. 2. Annotation process for sketch image

the predicted probabilities. We used the pretrained convolutional weights on ImageNet to train the object detection system for GUI elements by the way shown in YOLO official site [7]. Figure 3 shows the schematic diagram of an automatic code generation on sketch image. The object detection algorithm is used for detecting GUI elements in the first step.

3.3 Code Generation

After the finding graphic objects, annotation including bounding box and class information is transformed to hierarchical semantic structure. Since the bounding box information includes the numerical position information, we should extract the relative positional and sequence information. After the extraction of meaningful information, the meta information is translated to the given platform dependent GUI codes.

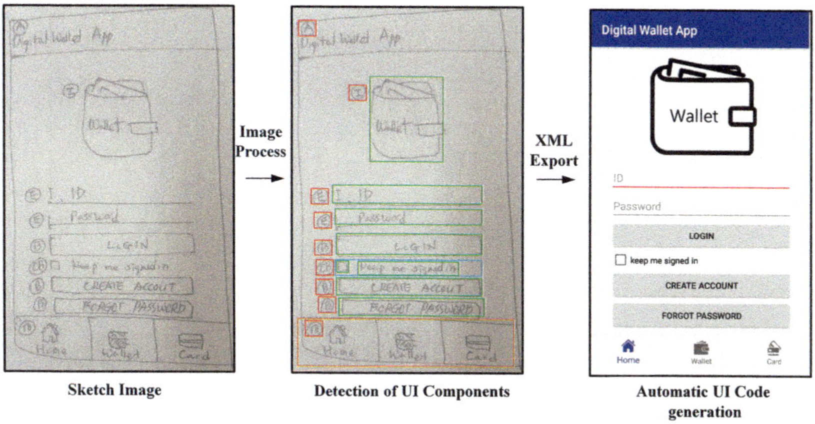

Fig. 3. Automatic GUI code generation scheme on sketch image

4 Work in Progress

We designed the automatic code generation system for GUI components which are manually drawn with pen on the paper. At first, the system recognizes the elements and describes the GUI elements with hierarchical object description language. These descriptions are parsed and translated to the target platform languages.

Acknowledgment. This research was supported by Next-Generation Information Computing Development Program through the National Research Foundation of Korea (NRF) funded by the Ministry of Science, ICT (No. NRF-2017M3C4A7069073).

References

1. Parag, T., Bahlmann, C., Shet, V., Singh, M.: A grammar for hierarchical object descriptions in logic programs. In: 2012 IEEE Computer Society Conference on Computer Vision and Pattern Recognition Workshops, Providence, RI, pp. 33–38 (2012)
2. Pham, H., Nguyen, T., Vu, P., Nguyenm, T.: Toward mining visual log of software. arXiv e-prints arXiv:1610.08911, October 2016
3. Beltramelli, T.: pix2code: generating code from a graphical user interface screenshot. arXiv e-prints arXiv:1705.07962, June 2017
4. Girshick, R., Donahue, J., Darrell, T., Malik, J.: Rich feature hierarchies for accurate object detection and semantic segmentation. In: 2014 IEEE Conference on Computer Vision and Pattern Recognition, Columbus, OH, pp. 580–587 (2014)
5. Redmon, J., Farhadi, A.: YOLO9000: better, faster, stronger. arXiv e-prints arXiv:1612.08242v1, December 2016
6. Tzutalin: LabelImg, Git code (2015). https://github.com/tzutalin/labelImg
7. YOLO: real-time object detection. https://pjreddie.com/darknet/yolo/

Emotion and Fatigue Monitoring Using Wearable Devices

Jong-Seok Lee[1], You-Suk Bae[2], Wongok Lee[2], Hyunsuk Lee[2(✉)], Jinkeun Yu[2], and Jong-Pil Choi[3]

[1] School of Integrated Technology Yonsei University, 85, Songdogwahak-ro, Yeonsu-gu, Incheon, Korea
jong-seok.lee@yonsei.ac.kr

[2] Clart Co., LTD, 250, Hagui-ro, Dongan-gu, Anyang-si, Gyeonggi-do, Korea
{ysbae,goks.lee,leehs,yjk760}@clart.kr

[3] Department of Computer Engineering, Korea Polytechnic University, 237, Sangidaehak-ro, Siheung-si, Gyeonggi-do, Korea
jpchoi@kpu.ac.kr

Abstract. This paper presents our prototype wearable system for monitoring emotion and fatigue of users. We develop a hardware part that can measure Galvanic skin response and photoplethysmography at a sampling rate of 100 Hz. In addition, we build a classification module that can distinguish the type of emotion and the level of fatigue of the user based on the measured signals. It is demonstrated that the developed system can successfully be used for emotion and fatigue monitoring.

Keywords: Galvanic skin response · Photoplethysmography · Classification Neural networks

1 Introduction

With the advances in wearable devices such as smart bands, it has become possible to constantly monitor users' mental and physical states and provide feedbacks based on the sensed states [1]. In particular, physiological channels such as heart rate, Galvanic skin response (GSR), and skin temperature provide useful information regarding the user's state. Promising applications include health monitoring, stress management, content recommendation, etc. For instance, if the level of fatigue detected by a smart band is too high during driving, some rest or sleep can be suggested; if the user is found to be depressed by a smart band, some joyful movies can be recommended to change the user's mood.

In this paper, we concentrate on developing a prototype wearable system for emotion and fatigue monitoring. It measures the GSR and photoplethysmography (PPG) signals of the user and perform recognition of the emotion type and level of fatigue. We develop a hardware prototype equipped with the sensors. And, we design a software-based recognition part that analyzes the signals and conducts classification using neural networks.

S. O. Hwang et al. (Eds.): ICGHIT 2018, LNEE 502, pp. 91–96, 2019.
https://doi.org/10.1007/978-981-13-0311-1_17

The remainder of the paper is organized as follows. Section 2 describes the hardware prototype. In Sect. 3, the recognition system is explained including data collection, feature extraction, and classifier design. The experimental results are presented in Sect. 4. Finally, conclusion is given in Sect. 5.

2 Hardware

We develop a prototype circuit board that can be integrated in a wearable device to measure physiological responses of the user. In particular, a GSR sensor module and a PPG sensor module are included. Besides, our prototype also equips a microcontroller unit (MCU), a power supply module, a USB communication module, and a Bluetooth communication module. A maximum sampling frequency of 1000 Hz is supported.

Figure 1 shows a user's hand connected with the developed sensor board. In Fig. 2, example GSR and PPG signals collected using the board are shown.

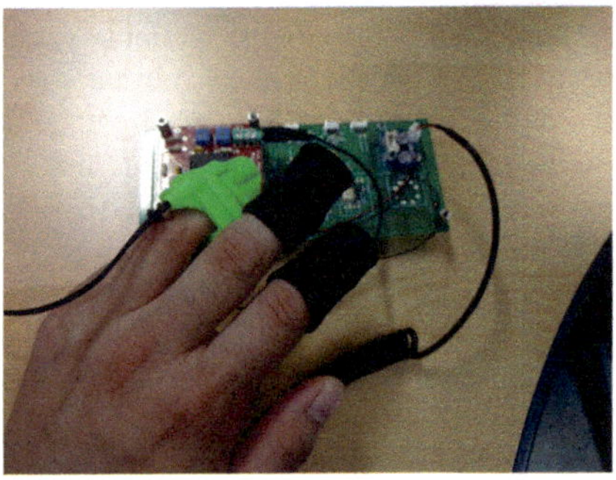

Fig. 1. Developed system attached to a hand.

3 Recognition System

3.1 Data Collection

Ten subjects were employed for data collection. Each subject took part in eight sessions held in different days. In each session, the experimental procedure is as follows. First, the objective and procedure of the experiment were explained. Then, the GSR and PPG sensors were placed at a hand of the subject sitting on a chair. The subject was instructed to relax for five minutes, during which the GSR and PPG signals for the resting state were recorded. After an additional rest period of three minutes, three music videos were played to induce particular emotional states, i.e., joy, anger, and sadness, respectively, during which the GSR and PPG signals were recorded. After the end of a music video and before the start of another, a break for three minutes was given to allow the subject's emotional state to revert to a normal state. The subject wore an earphone to listen to music, so that any environmental noise did not interrupt the

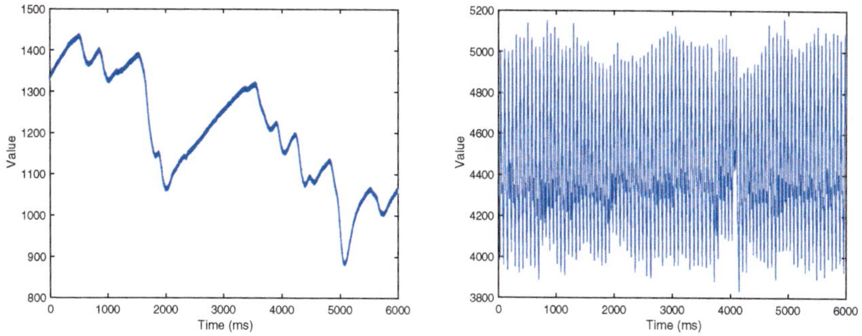

Fig. 2. Example GSR (left) and PPG (right).

subject's immersion to the stimuli and induction of emotional arousal. In the whole experiment, a specific music video was shown only one time.

In each session, the subject was asked to rate the level of fatigue in the Likert scale ranging from 1 (not tired at all) to 9 (extremely tired).

The physiological signals were recorded at a sampling frequency of 100 Hz.

3.2 Feature Extraction

First, the original signals were down-sampled to 10 Hz. Then, each signal was divided into segments by using a window having a length of 60 s and moving 5 s at a time. Each segment becomes a data point in classification.

For each segment, we extract seven features from GSR, and five features from PPG, which have been known to be effective for physiological signal analysis [2–4]. It should be noted that in literature, there are abundant types of features that can be extracted from GSR and PPG, but due to the limit set by the hardware specification, we had to choose the most effective features among them.

The seven GSR features are as follows:

- Average value of the signal
- Standard deviation value of the signal
- Average value of the derivative of the signal
- Standard deviation value of the derivative of the signal
- Average value of the signal after low-pass filtering with a cut-off frequency of 0.2 Hz
- Average value of the derivative of the signal after low-pass filtering with a cut-off frequency of 0.2 Hz
- Number of peaks in the signal after low-pass filtering with a cut-off frequency of 0.2 Hz

The five PPG features are as follows:

- Average height of the peaks in the signal
- Standard deviation value of the height of the peaks in the signal
- Heart rate

- Standard deviation value of the peak-to-peak intervals
- Average value of the squared heart rate variability

3.3 Classification

The emotion classification task is defined to recognize the given signal as one of the four classes, i.e., neutralness, joy, anger, and sadness. The four classes of the fatigue level classification task correspond to 1 and 2, 3 and 4, 5 and 6, and 7 and 8 in the rating scale. Note that score 9 was never given by the subjects, therefore it was not considered in the classification.

Two classification schemes are tested, namely, the subject-wise scheme and the subject-dependent scheme. In the subject-wise scheme, a separate classifier is built for each subject, where the first half of the signal is used for training and the remaining half for testing. The average classification performance over all subjects is reported. In the subject-dependent scheme, a classifier is constructed using the first half of the signals of all subjects and tested using the remaining data of all subjects.

We use multilayer neural networks having one hidden layers as classifiers. We try various numbers of hidden neurons to examine the performance with respect to the neural network complexity. A sigmoid function is used as the activation function of each hidden neuron. The neural networks are trained using the Levenberg-Marquardt algorithm that is one of the fastest neural network training algorithms. The maximum training epoch is set to 200.

4 Results

Figure 3 shows the classification performance for the subject-wise scheme. The emotion classification accuracy reaches the best (61.9%) when 50 hidden neurons are used, while the fatigue classification accuracy is the highest (88.2%) when the number of hidden neurons is 40. It is observed that the performance becomes saturated once a sufficient number of hidden neurons is used.

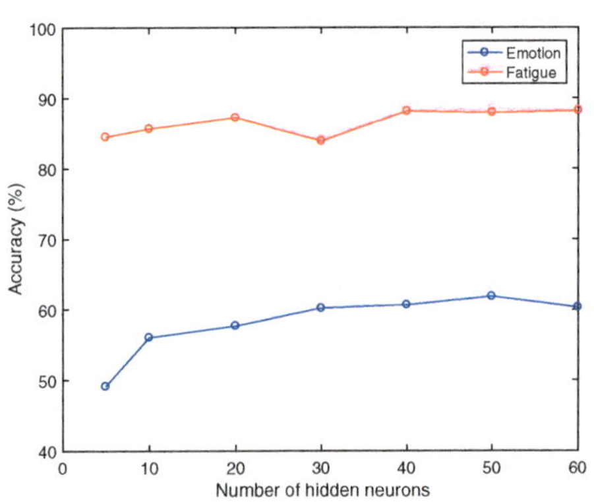

Fig. 3. Performance of subject-wise classification for emotion and fatigue

In Fig. 4, subject-dependent classification performance is depicted. It is observed that the accuracy in this figure is lower than that in Fig. 3, which is due to the subject-wise variation of the patterns appearing in the physiological signal. The best performance is obtained when the number of hidden neurons is 50 for emotion (48.4%) and 50 for fatigue (71.2%).

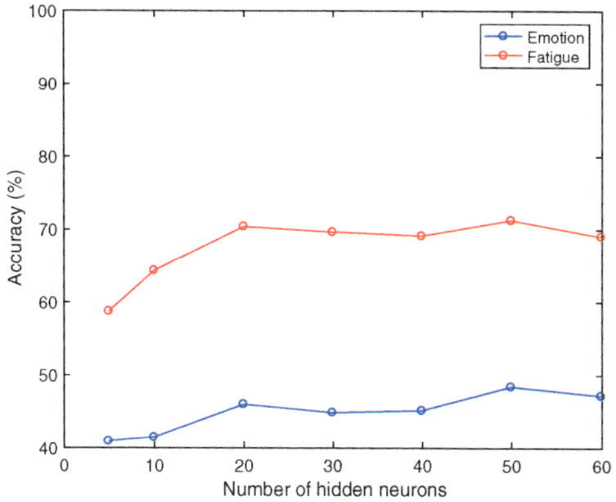

Fig. 4. Performance of subject-dependent classification for emotion and fatigue

5 Conclusion

In this paper, we have presented our prototype system for emotion and fatigue monitoring. Through an experiment, it was shown that the hardware and software parts can monitor the type of emotion and the level of fatigue with a satisfactory accuracy.

The results showed that user-dependent variation can be a source of performance degradation. In the future, therefore, it will be desirable to focus on reducing the effect of such user-dependent variation in order to make the system work robustly across different users.

Acknowledgment. This work was supported by the Technology Innovation Program (10052841, "Development of stretchable smart band for young people") funded by the Ministry of Trade, Industry and Energy, Korea.

References

1. Moon, S.-E., Lee, J.-S.: Implicit analysis of perceptual multimedia experience based on physiological response: a review. IEEE Trans. Multimedia **19**(2), 340–353 (2017)
2. Alberdi, A., Aztiria, A., Basarab, A.: Towards an automatic early stress recognition system for office environments based on multimodal measurements: a survey. J. Biomed. Inform. **59**, 49–75 (2016)

3. Healey, J.A., Picard, R.W.: Detecting stress during real-world driving tasks using physiological sensors. IEEE Trans. Intell. Transp. Syst. **6**(2), 156–166 (2005)
4. Koelstra, S., Muehl, C., Soleymani, M., Lee, J.-S., Yazdani, A., Ebrahimi, T., Pun, T., Nijholt, A., Patras, I.: DEAP: a database for emotion analysis using physiological signals. IEEE Trans. Affective Comput. **3**(1), 18–31 (2012)

Process and Evaluation Index of Bankart Surgery for the Simulator Development

Min-Jae Lee(✉), Ein Jeong Hwang(✉), and Do Hoon Oh(✉)

Institute for IT Convergence, Myongji Hospital, Goyang-si, Gyeonggi-do, Korea
alexlmj90@gmail.com, einjeong@gmail.com,
dohoon5@yahoo.co.kr

Abstract. This paper's purpose is to analyzing the process of Bankart. The reason to analyze this process is that there is various method of the surgery each professional surgeon. So, we have to make a standard of the Bankart Surgery. And we have studied about arthroscopic surgery such as Bankart and Rotator cuff to develop medical simulator for resident education. Because arthroscopic surgery is difficult and complex, it is not easy to learn surgery in the book and they need practical experience as well as theoretical experience. If we complete development of simulator, it helps resident to learn feeling of the 'Bankart' and 'Rotate cuff' surgery. Residents who want to be a professional doctor can't do better than using medical simulator. To give them a detail and actual information, we have classified the process of Bankart and sorted evaluation index in this paper. That's why to analyze the process of Bankart surgery in this paper.

Keywords: Surgery · Education · Bankart repair · Rotator cuff Shoulder · Simulator · Evaluation index · Arthroscopy · VR

1 Introduction

1.1 Words About Shoulder and Surgery

For understanding this paper, we have to know some of words about body part and understand what is their function.

- Joint pit: A concave surface of two bones forming joints.
- Humerus: A bone that extends from the shoulder to the elbow.
- Biceps: A large muscle in the front part of the upper arm.

Shoulder tissues are composed of important tissue and bone. So this muscle and tissue can rotate our arm. In this paper, there are three important word were explained (Fig. 1).

Humerus
It forms the shoulder joint and the arthritic joint.
The shoulder joint produces six exercise functions consisting of flexion, extension, adduction, abduction, internal rotation and external rotation.
In the joints of the arthritic joints, four exercise functions are created, consisting of flexion, extension, pronation (1), and supination (2).

S. O. Hwang et al. (Eds.): ICGHIT 2018, LNEE 502, pp. 97–102, 2019.
https://doi.org/10.1007/978-981-13-0311-1_18

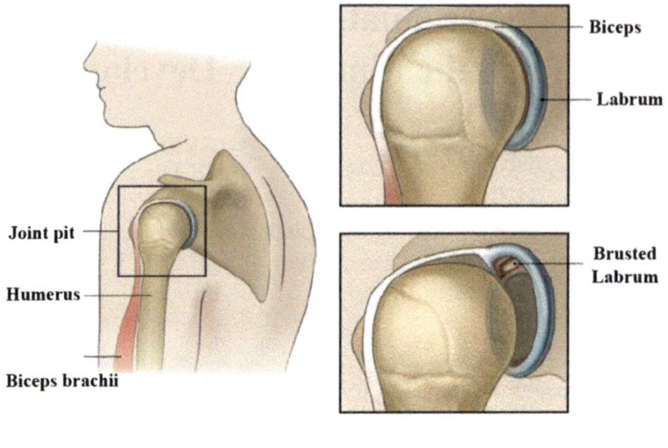

Fig. 1. Explanation of Labral tear.

Biceps brachii
It functions as bending muscles of the lower arm when the lower arm is in the back position. A short one prevents the dislocation of the shoulder joint.

Labrum
It is a fibrous cartilage that surrounds the joints and edges of the scapula from the bones that form the shoulder joints.

A Bankart Shoulder Repair
It is a surgical technique for the repair of recurrent shoulder joint dislocations.

1.2 Explanation About the Bankart

A Bankart Shoulder Repair is a surgical technique for the repair of recurrent shoulder joint dislocations. Within the procedure, the worn out ligaments are re-attached to the proper place in the shoulder joint, using the objective of rebuilding normal function. The shoulder is easily the most commonly dislocated major joint following severe trauma, such as an auto collision or perhaps a fall onto an outstretched arm. Some 96% of dislocations involve the front of the shoulder (anterior), with 1–3% occurring in the back (posterior).

(1) pronation: inner rotate (2) supination: outer rotate (Fig. 2).

Falls and car accidents are common causes of first-time dislocations, but recurrent dislocations in many cases are because of seemingly inoffensive activities such as raising the arm over the head, or combing hair. Shoulder dislocations are more common in males than females, as well as in teenagers rupture can be divided into the case of rupture into the partial layer and the case of rupture into the whole layer.

In particular, the upper joints and the sutures are linked to the biceps in the arms and loosely attached to the bones like the meniscus of the knee joints.

So, it is prone to be relatively damaged.

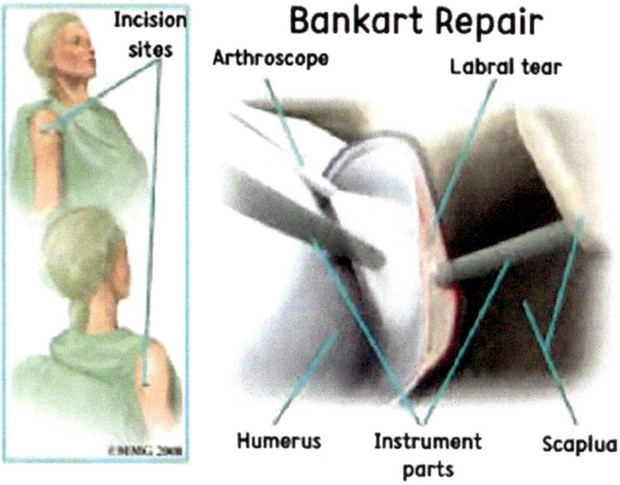

Fig. 2. Bankart shoulder repair for unstable dislocating shoulders

Especially, Bankart lesion is a rupture of the anterior and posteromedial joints, which means the loss of function of the lower and upper brachial ligaments which play the most important role in joint stability.

2 Research Way

2.1 The Step of Research

First, Search the Web and Books

We had to search about the surgery through books and Internet.

In the book, we could tell the difference between the past surgery method and the current surgery method and process. On the internet, we had known to the surgery methods and opinions of other surgery. It solved the problem of limited information that we could only know by a doctor.

Second, Observe the Surgery

We had to observe Bankart and Rotate cuff surgery from surgeon. We realized that the surgical procedure may be changed depending on the patient's condition. We need to record what is the difference of each surgery.

Third, Collect and Analyze Data

We collect and analyze surgery data about Bankart surgery. We determined the evaluation index by recording the appropriate surgical procedure and easy-to-mistake procedure for the patient.

3 Main Subject

3.1 Process of Bankart Surgery

The medical simulator is developed to enable experienced practitioners or practitioners who have little experience in training to directly experience similar operations or procedures.

In addition to the definitions and methods of existing medical simulators, evaluation indicators and reference points, which are essential elements for judging this, and correct surgical procedures are required.

First, we divided process to 4 parts in the Bankart surgery (Fig. 3).

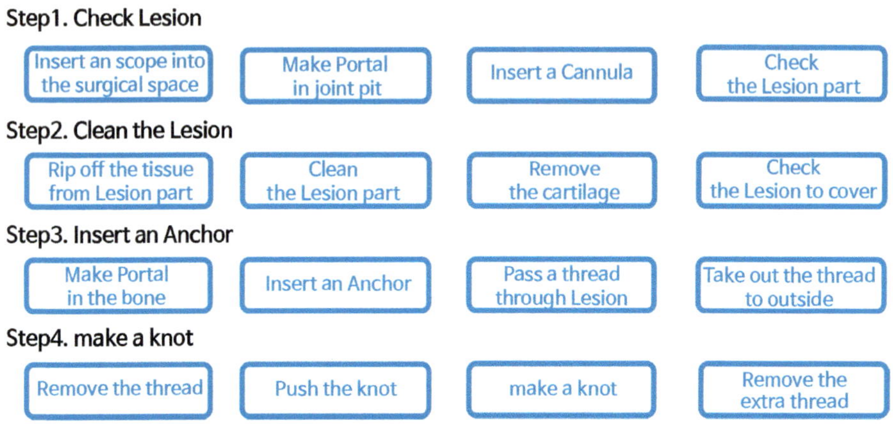

Fig. 3. Process of bankart shoulder repair

3.2 Evaluation Indexes

Step 1, Check Lesion

First, Surgeon has to insert a scope into the posterior portal.

[Evaluation index 1] is estimation of the distance to acromion from coracoid.
[Evaluation index 2] is that finding exactly soft spot and inserting scope smoothly.
[Evaluation index 3] is that making a portal well to insert other surgical tools.
[Evaluation index 4] is that spending time to insert needle, mass and other surgical tools. Next, he inserts cannula and check the lesion part by using probe.

Step2, Clean the Lesion

Surgeon has to rip off the tissue from lesion part for clear view by using osteotome (1).

[Evaluation index 1] are sight to see clearly, and time to spend and he has to remove Bursa by using Shaver then, take care of the blooding. And then bleed the tissue by using electrocautery (2).
[Evaluation index 2] are ratio of Bursa in joint pit, and time to spend.
[Evaluation index 3] is that amount of blood on the tissue.

Step 3, Insert an Anchor
Surgeon has to make a hole in the bone by using Awl.

[Evaluation index 1] is that after taking off Awl, inserting Anchor to bone until mark by using mallet. He has to pass the thread through lesion.
[Evaluation index 2] is estimation of spending time to take out the thread to outside by using Retriever.
[Evaluation index 3] is using Hook, evaluation of pulling the thread through the ring (Fig. 4).

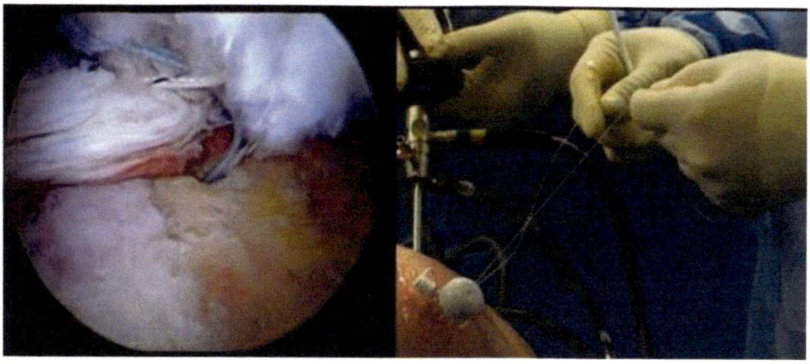

Fig. 4. Inserting anchor to patient in bankart repair

Step 4, Make a knot
Surgeon has to Remove the thread and push a knot.

[Evaluation index 1] is whether the knot is pushed out from the outside and the knot is not twisted to the lesion. He makes a knot repeatedly 2–3 and then cut the knot. Lastly, Remove the extra thread.
[Evaluation index 2] is how to spend time to working and degree of twisted thread. After We have observed some of Bankart surgery, we could find a common things and different things. And then analyzing process and evaluation index.
(1) osteotome: surgical tools to clean the entangled tissue (2) electrocautery: surgical tools used for hemostasis.

4 Result

4.1 Current Achievements and Future Direction

We have studied the process and evaluation index of the surgeons. And this study will be used to help residents learn similar real experiences (Fig. 5).

Currently we are in the process of making a simulator machine and plan to apply the evaluation index here. Through this study, we will provide a variety of surgical environments for residents to help with education.

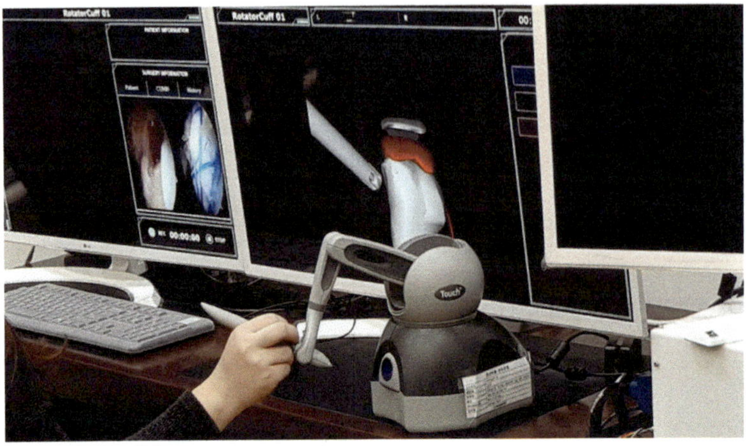

Fig. 5. Prototype of simulator machine

Acknowledgments. This submission was carried out with support from the research fund of Project for the development of simulator Virtual Surgery Simulator Technology Development for Medical Training. (Project No: 10053260).

References

1. Asan Medical Cente Hompage. http://www.amc.seoul.kr/asan/healthinfo/disease/diseaseDetail.do?contentId=33889. Accessed 27 May 2015
2. St. George – surgical center. https://www.sgsc.net/procedure/Bankart-repair-unstable-dislocating-shoulders/. Accessed May 2015
3. Naviworks, simulator Virtual Surgery Simulator Technology Development for Medical Training: Image and Process Analysis for Virtual Shoulder Arthroscopy (2018)
4. https://blog.naver.com/jbpark2/220694859204. Accessed 27 Apr 2016
5. Arthrex, Bankart and Remplissage Repair. https://www.arthrex.com/resources/video/btfmfT15506jwgFf23Vfhw/Bankart-and-remplissage-repair. Accessed 21 Nov 2017

Application of Use Cases for Congestive Heart Failure

Hyo Seon Kim(✉), Ein Jeong Hwang(✉), Sam Lee(✉), Deok Kyu Cho(✉), and Do Hoon Oh(✉)

Institute for IT Convergence, Myongji Hospital, Goyang si, Gyeonggi do, Korea
khs26195131@gmail.com, einjeong@gmail.com, chodk123@gmail.com, cutegirl11526@naver.com, dohoon5@yahoo.co.kr

Abstract. This paper is about use cases of congestive heart failure. The purpose of this paper is to know through the use of the congestive heart failure use cases the currently developing smart patch at some stage to apply to patients with congestive heart failure. This smart patch has a total of three sensors. The sensor is an electrocardiogram, acceleration, impedance. We examined definitions, symptoms, causes, diagnosis, and methods of congestive heart failure and we have described the process from the stage where the patient first visited the hospital to the stage where he or she is discharged to the hospital, as a total of ten stages of use cases. The smart patch currently under development can be applied at the stage of confirming the therapeutic effect of congestive heart failure patients.

Keywords: Congestive heart failure · Smart patch · Impedance sensor

1 Introduction

The number of patients with heart failure has steadily increased over the years. The number of heart failure patients increased from 116,085 in 2013 to 122,553 in 2016 [1]. Professor Min Seok Kim of Ulsan Medical University (Department of Cardiology, Seoul Asian Hospital) said, "According to the survey findings of the domestic heart failure registration, the average age of patients with heart failure is increasing according to the aging of the population, "Heart failure is a social economic burden It is necessary to have interest in treatment and management of heart failure in that it is [2]." Also as shown in Table 1, for heart disease, cancer causing death in Korea 10 in 2016 occupied the second place [3].

Table 1. Top 10 death causes and mortality rates, 2016 (People per 100,000 people)

Ranking	Cause of death	Death rate	15 year rank comparison
1	Malignant neoplasia (cancer)	153.0	-
2	Heart disease	58.2	-
3	Cerebrovascular disease	45.8	-

S. O. Hwang et al. (Eds.): ICGHIT 2018, LNEE 502, pp. 103–108, 2019.
https://doi.org/10.1007/978-981-13-0311-1_19

2 Smart Patch

Currently, we are developing two kinds of smart patches. Figure 1 shows the multiple bio-signal patch. The multiple bio-signal patch in it measures the electrocardiography, acceleration, weight. Therefore, for congestive heart failure patients, measuring body weight is very important as it is a means to confirm the effect of treatment. We will find a procedure in which patients with congestive heart failure can be applied to patients with congestive heart failure by making the use cases from the first hospitalization until discharge until discharge.

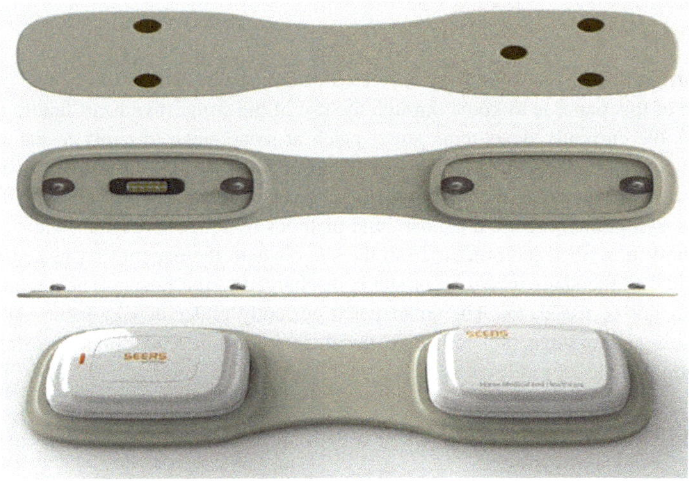

Fig. 1. The multiple bio-signal patch

3 Congestive Heart Failure

3.1 Definition

Generally speaking, heart diseases causing congestion in whole body Venensystem due to pump impairment of the heart can't satisfy metabolic demands of body tissues to reduce cardiac output: It is not a diagnosis name of a disease but a pathophysiological clinical condition as a result of heart disease [4].

3.2 Factors of Cause and Deterioration

Primary heart failure inducing factors are pulmonary diseases such as heart disease (myocardial infarction, angina, valvular heart disease, inflammatory heart disease, arrhythmia etc.) and COPD, and secondary heart failure inducing factors are anemia, infection thyrotoxicosis, Hypothyroidism, Paget's disease, nutritional deficiency, hyperhidrosis etc. Risk factors for heart failure are atherosclerosis, diabetes, hyperlipidemia, hypertension, metabolic syndrome, smoking, obesity etc. [4].

3.3 Symptoms

Symptoms of left ventricular failure are as follows. Blood is refluxed into the pulmonary vein and capillary, resulting in pulmonary congestion. Dyspnea, cough, breathing difficulty during exercise, paroxysmal nighttime breathing difficulty, comfortable breathing rather than sitting cold sitting comfortable sitting respiratory symptoms, symptoms such as pulmonary edema are displayed. This makes it easier to become fatigued and cause insomnia sometimes. Symptoms of right ventricular failure are as follows. Increases in pulse rate, general vein and capillary congestion symptoms and signs are displayed [5].

3.4 Diagnosis Method

Methods for diagnosing congestive heart failure include cardiac ultrasound, blood test, electrocardiogram, chest x-ray, coronary angiography etc.

4 Use Cases for Congestive Heart Failure

4.1 Visit to Patient Hospital

Figure 2 shows the use case of the process from the path where patients with congestive heart failure will visit to the hospital first until the patient is discharged. The first steps are as follows. In the case of patients with acute heart failure, severe breathing difficulties, coughing symptoms appear, symptoms of coughing appear in the sputum, vivid red blood is mixed out, lips and fingertips are made blue, patients with cold sweat cannot also lie sideways. If proper treatment is not initiated, the patient can die and need to move immediately to a hospital where first aid is available. In the case of patients with chronic congestive heart failure, they will visit through the outpatient,

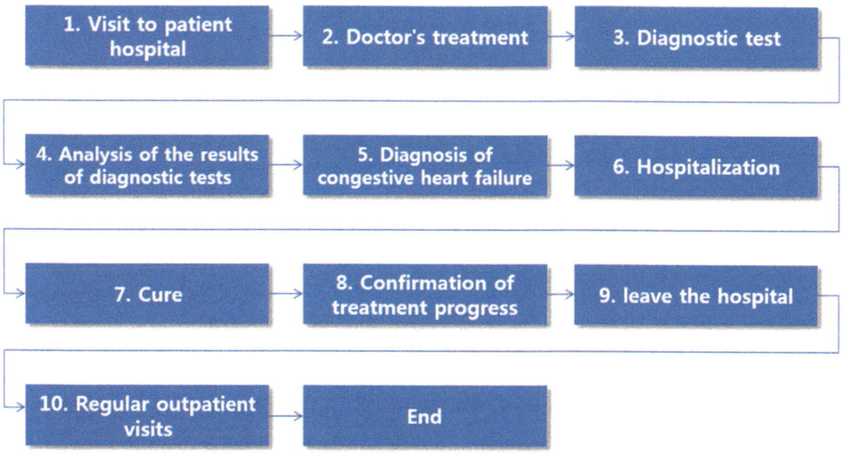

Fig. 2. Use cases for congestive heart failure

complain that breathing difficulties and edema are accompanied and the feeling of fatigue gradually becomes severe [6].

4.2 Doctor's Treatment

The doctor will check the symptoms of the patient. Then presume congestive heart failure and prescribe an appropriate test for it.

4.3 Diagnostic Test and Analysis of the Results of Diagnostic Tests

Heart failure can be diagnosed by integrating clinical symptoms, physical examinations and electrocardiograms, and chest X-ray. In order to establish heart failure and make a treatment plan, it is necessary to perform ultrasonic examination of the heart. In addition to confirming heart failure by cardiac ultrasound examination, in many cases, cause disease can also be found. Figure 3 shows a cardiac ultrasound examination. In addition to the cardiac ultrasound examination method, there are blood test, coronary angiography, MRI, cardiac tissue examination etc. in the examination method to find the cause of heart failure. Medical personnel and medical technicians analyze the results of the above diagnostic method and report the doctor in charge [7].

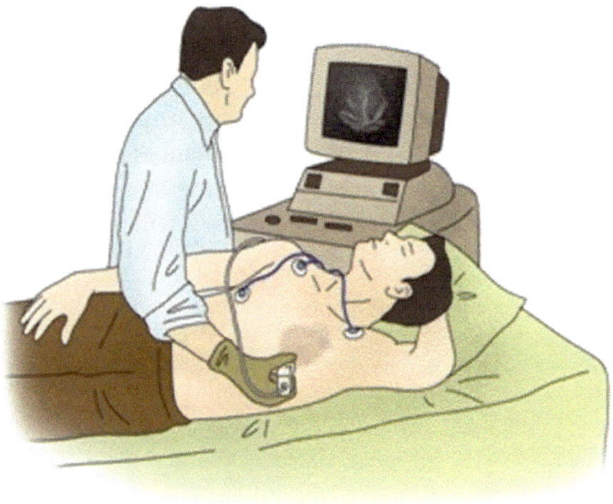

Fig. 3. Cardiac ultrasound examination

4.4 Diagnosis of Congestive Heart Failure and Hospitalization

Through the analyzed results, the doctor diagnoses that the patient is congestive heart failure and prescribes the patient's hospitalization. And the patient is hospitalized.

4.5 Cure

The goal of treatment is to eliminate body fluid accumulated excessively in the body, increase myocardial contractility and efficiency, and reduce the workload of the heart. Use a diuretic to remove excess body fluid, reduce ventricular pressure, limits low salt diet and water intake. Also, use drugs to improve myocardial contractility, relax peripheral blood vessels, etc. Reduce the burden on the heart Treatment to increase cardiac output to treat. In case of severe heart failure done, heart transplant surgery will be performed [5].

4.6 Confirmation of Treatment Progress

Medical personnel measure the patient's weight each morning in order to grasp the influence of the therapeutic drug (diuretic, tonic, etc.) administered to the patient on the body and evaluate the calories, moisture amount, excretion amount accompanying it, and the amount of calories ingested by the patient on a daily basis, and confirm that the ratio is within 0 ml at −1000 ml [8].

4.7 Leave the Hospital

The patient may be discharged when the medication injected by the injection is convertible into oral medication

4.8 Regular Outpatient Visits

Patients should take the prescribed medication well and visit hospital regularly, smoking, abstinence from drinking, blood pressure control, and stress control, using low-salt diets and exercise.

5 Conclusion

Currently, we are developing multiple bio-signal patch. The objective of this patch development is to develop a low power consumption, small patch type device for measuring biological signals, monitor cardiovascular diseases, and make prognosis management possible. This paper summarized the use cases of the process from the path where patients with congestive heart failure will visit to the hospital first until the patient is discharged. We investigated which stage of use cases of congestive heart failure, which is one of cardiovascular diseases, patch of multiple bio-signal of electrocardiogram sensor, acceleration sensor, Impedance sensor can be applied. According to the confirmation procedure of the course of treatment of the use cases, medical personnel measure the patient's weight every morning in order to confirm the course of treatment of patients with congestive heart failure. Further, the impedance sensor can measure the weight of the patient through impedance variation analysis. Therefore, through many clinical trials, we will confirm whether smart patch can be applied to patients with congestive heart failure.

Acknowledgements. The research and writing of this article were made possible by funding from the "Development of human-body-friend smart patch and home-based healthcare service solution for monitoring acute/chronic brain/cardiovascular diseases (No. 10053091)".

References

1. Health Insurance Review & Assessment Service Homepage. http://opendata.hira.or.kr/op/opc/olap3thDsInfo.do. Accessed 07 Dec 2017
2. Medical Observer Homepage. http://www.monews.co.kr/news/articleView.html?idxno=107877. Accessed 20 Nov 2017
3. Statistics Korea Homepage, Press release on 2016 death cause statistics. http://kostat.go.kr. Accessed 22 Sept 2017
4. Korean Nurse Association: The great Encyclopedia of Nursing Science. Korean Dictionary Researcher, Seoul (1996)
5. Samsung Medical Center Homepage. http://www.samsunghospital.com/home/healthInfo/content/contenView.do?CONT_SRC_ID=09a4727a8000f2ed&CONT_SRC=CMS&CONT_ID=2438&CONT_CLS_CD=001020001004. Accessed 05 F2014/02/05
6. Korea University Medical Center Homepage. http://anam.kumc.or.kr/dept/disease/deptDiseaseInfoView.do?BNO=127&cPage=2&DP_CODE=AACA&MENU_ID=004004. Accessed 2012
7. Asan Medical Center Homepage, http://www.amc.seoul.kr/asan/healthinfo/disease/diseaseDetail.do?contentId=30314. Accessed 25 May 2014
8. Konkuk University Chungju Hospital Homepage, Patient Nursing Guidelines. http://www.kuh.co.kr. Accessed 2006
9. KIMS Co., Ltd. Homepage. https://www.kimsonline.co.kr/ResCenter/diseasefocus/view/150. Accessed 2011
10. Sik, K.D., Won, K.B., Chul, C.S., et al.: Clinical Physiology (Electrocardiogram, Cardiac Sound). Korea Medical Book Publishing Company, Seoul (2012)
11. Yeong, P.S., Seon, P.Y., Jun, B.H., et al.: Clinical Physiology (Echocardiography, Cerebral Blood Flow Test). Korea Medical Book Publishing Company, Seoul (2012)

Epileptic Seizure Detection Using Empirical Mode Decomposition Based Fuzzy Entropy and Support Vector Machine

Deepti Tripathi and Navneet Agrawal(✉)

Department of ECE, CTAE, MPUAT, Udaipur, Rajasthan, India
deeptitripathi08@gmail.com, navneetctae@gmail.com

Abstract. A neurological condition affecting the central nerve system of people causing recurring seizure is termed as epilepsy or seizure disorder. A seizure can be described as a brief, interim disturbance in the electrical activity of the brain and the cause of its occurrence is the quick firing of many nerve cells in the brain, causing an electrical storm. Almost 50 million people worldwide have epilepsy and its studies often rely on EEG signals for analyzing brains behavior during the occurrence of seizure. Many kinds of research around the world is carried out over past few years to analyze EEG signals automatically to detect epilepsy and type of seizure present. This paper presents classification of EEG signals into healthy/inter-ictal versus ictal EEGs using EMD based fuzzy entropy method and SVM classifier. In EMD, decomposition of the EEG signal from different epileptic states takes place to obtain IMFs. Fuzzy entropy reduces the detection time by reducing the data size of the EEG data without any loss of the information. So here EMD is followed by the calculation of FuzzyEn of the reconstructed signal obtained from IMFs. Finally, feature vectors are formed using fuzzy entropy that are applied to non-linear SVM classifier for classification purpose. The classification accuracies obtained by using proposed method for Z versus S, O versus S, N versus S, and F versus S when only 50% data is applied during testing of the classifier are 99.88%, 99.38%, 98.62%, and 97% respectively.

Keywords: EEG · EMD · Fuzzy · SVM

1 Introduction

Epilepsy can be described an event in which group of nerve cells signals abnormally, disturbing the normal pattern of the neuronal activity, and this disturbance causes unusual sensations or loss of consciousness. During a seizure, neurons may fire much faster than that in normal conditions as many as 500 times a second. It may develop due to an imbalance of nerve signaling chemicals. Seizure and epilepsy are not always same, that is seizure can be described as short term disturbance in the brain's electrical activity where as such recurring disturbances is epilepsy.

Epileptogenesis is the gradual process by which a normal brain develops epilepsy. Brain activity in the seizure state (ictal) differs considerably from that in the normal state in terms of neuronal firing patterns and frequency. Seizure detection at an early

S. O. Hwang et al. (Eds.): ICGHIT 2018, LNEE 502, pp. 109–118, 2019.
https://doi.org/10.1007/978-981-13-0311-1_20

stage may allow precautions to be taken to avoid severe damage to the patient's brain, moreover once a seizure is fully developed precise information about the location is usually unavailable [1]. Among many methods available for diagnostic, EEG is considered as an effective technique in clinical treatments and research.

In 1920s, using scalp electrodes practical values of EEG was demonstrated by Berger [2]. Extra-cranial EEG provides information of the electro-cerebral activity of the entire brain whereas intracranial EEG presents information of the specific region of the brain where the electrodes are been surgically implanted.

A lot of work is being practiced in detection of epileptic seizure. Due to non-stationary nature of EEG signals many research groups have adopted nonlinear analysis. Lyapunov exponent [3], correlation dimension [4], fractal dimension [5], and entropy analysis [6] are among few of the nonlinear methods for extracting parameters that are relevant to the electrical activity of the brain. Non-linear entropy method helps in measuring the complexity involved in EEG signal, sample entropy (SampEn) [7], approximate entropy (ApEn) [8], permutation entropy (PE) [9], wavelet entropy [10], Shannon entropy [11] etc. are few works. Chen et al. [12] modified SampEn based algorithm and proposed the Fuzzy-En algorithm, which is a nonlinear index. To eliminate superimposed trends present in the EEG signals, that are responsible for increasing the standard deviation of the data and thus influencing entropy based analysis, empirical mode decomposition (EMD) is used for extracting intrinsic mode functions (IMFs) from the EEGs using [13].

The main aim of the present work is analyzing EEG signals by extracting the IMFs using EMD, and calculating FuzzyEn of the reconstructed signal. Later the FuzzyEn values are used to form feature vectors that are applied to our classifier, support vector machine (SVM), that distinguishes seizures signals from that of healthy person or seizure-free signals and at last accuracy, sensitivity and specificity of the method is determined.

This paper is structured in following manner: first, we have provided information about Bonn database used in present work in Sect. 2 and later block diagram of the proposed approach is explained in Sect. 3 followed by demonstration of the method adopted. Analysis of the work, results and conclusion are given in Sects. 4, 5 and 6 respectively.

2 Database

The present work is being practiced on Bonn database, consisting of five sets in which each set contains hundred single-channel EEG segments with duration and sampling rate of 23.6 s and 173.6 Hz respectively [14]. Z and O sets are segments recorded from the scalp of five healthy persons with their eyes open and closed respectively. Rests of the three sets are recorded using intracranial electrodes. Dataset F and dataset N delivers signals that were recorded from epilepto-genic zone and from hippocampal formation of the opposite hemisphere of the brain respectively. Both, F and N data sets, belongs to seizure free intervals, while as, set S contains seizure activity of the seizure patient.

3 Block Diagram and Method

In the present work, we extracted IMFs of the EEG data sets using EMD to eliminate trends present in the raw EEG data and calculated FuzzyEn values after signal reconstruction that acts as feature vectors thereof. These vectors are inputted to the SVM classifier for distinguishing the signals into normal, inter-ictal and ictal data sets. Block diagram of the proposed approach is shown in Fig. 1.

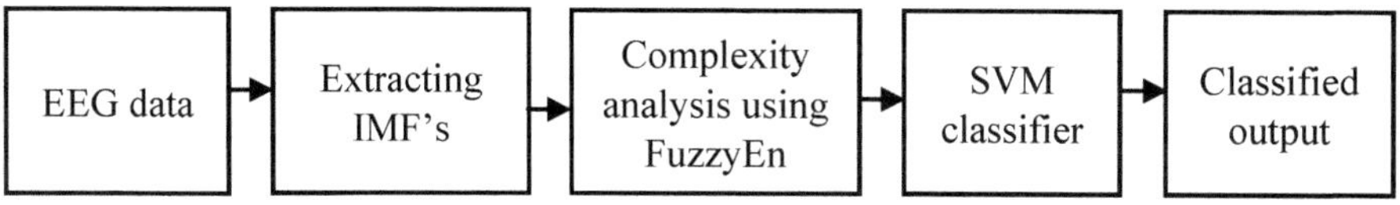

Fig. 1. The block diagram of proposed approach.

3.1 EMD Technique for De-trending Process

EMD technique decomposes any signal into IMFs that are actually a set of band-limited signals and satisfies two conditions, first is that the total number of extrema and zero crossings in the dataset must be either same or may differ atmost by one and the second condition is that the mean value of the envelope defined by the local maxima and that defined by local minima must be zero at any point. The algorithm that we used for EMD is precised as follows: First, we detected extrema of the signal x (t), i.e. F_{max} and F_{min} (both maxima and minima), and then used cubic spline interpolation to generate envelopes for maxima and minima as $e_{max}(t)$ and $e_{min}(t)$. This is followed by computation of mean as:

$$m(t) = \frac{(e_{max}(t) + e_{\min}(t))}{2} \tag{1}$$

The candidate of IMF is extracted as in (2) and checked whether d(t) in (2) is IMF or not. This is done by checking the two conditions described above, if yes d(t) is saved and the residue is computed as in (3).

$$d(t) = x(t) - m(t) \tag{2}$$

$$r(t) = x(t) - \sum\nolimits_{i=1}^{t} d(t) \tag{3}$$

At this stage incrementing t as t = t + 1, final residue is obtained by performing iteration on the residual till it becomes monotonic function, delivering no more IMF's. The reconstructed signal, $x'(t)$ is obtained by summing up saved d(t) as in (4). The original signal, after EMD process can be represented as summation of IMFs and the final residue and is presented in (5).

$$x'(t) = \sum_{i=1}^{m} d(t) \quad (4)$$

$$x(t) = r(t) + \sum_{i=1}^{m} d_i(t) \quad (5)$$

Here in (4) and (5) m is upper boundary number of the selected IMFs components. Figure 2 shows the entire process of EMD in a flow chart. Figure 3 shows seizures patients EEG signal and after applying EMD its IMFs and residue. The reconstructed signal obtained in (4) is applied to the next step of obtaining FuzzyEn values.

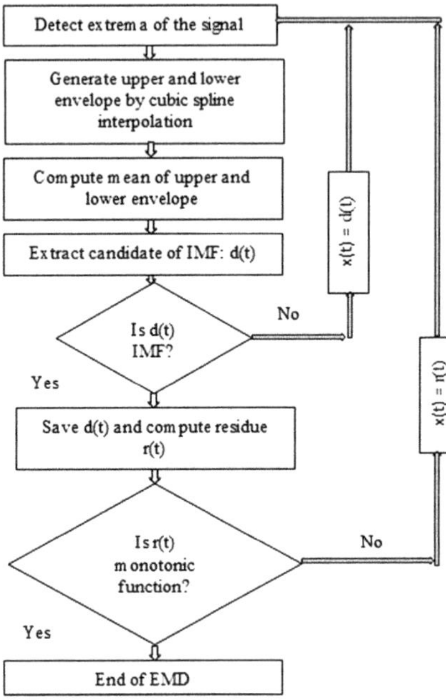

Fig. 2. Flow chart for EMD process.

3.2 Feature Selection Parameter

The entire series of all five sets i.e. Z, O, F, N and S after IMF extraction and signal reconstruction is divided into five different segment lengths of 173, 256, 512, 1024 and 2048 samples in each respectively; these segments are referred to as patterns. Value of FuzzyEn is calculated for each of these patterns separately.

3.3 FuzzyEn Algorithm for Complexity Evaluation

FuzzyEn is consistant with less dependency on data length [13] and its algorithm for time series $\{u(i): 1 \leq i \leq N\}$, with N number of sample values, can be explained as follows: First, we obtained vector sequence X_i^m as given in (6), where, i = 1, 2, …, N − m + 1, X_i^m represents m consecutive u values, commencing with the ith point and

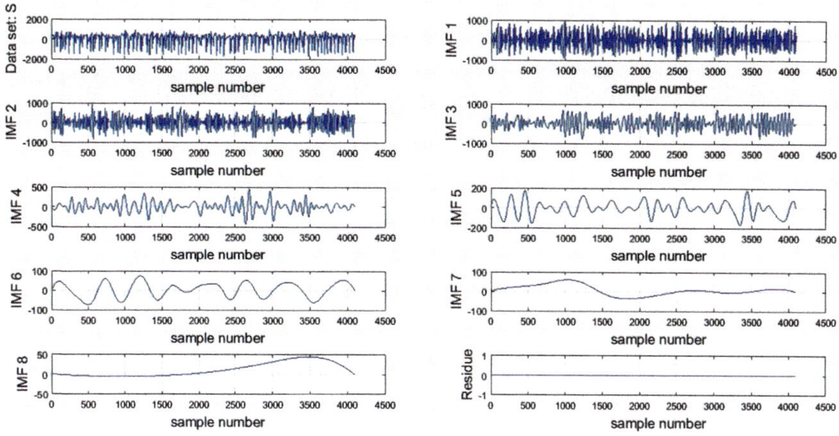

Fig. 3. Seizure data set with IMFs and residue.

m is length of the sequence to be compared. This vector is generalized by elimination of baseline $u_0(i)$ as given in (6) and defined in (7).

$$X_i^m = \{u(i), u(i+1), \ldots, u(i+m-1),\} - u_0(i) \tag{6}$$

$$u_0(i) = m^{-1} \sum\nolimits_{j=0}^{m-1} u(i+j) \tag{7}$$

Then similarity degree, D_{ij}^m between X_i^m and X_j^m was computed by us using fuzzy membership function as in (8).

$$D_{ij}^m = \exp\left(-\frac{d_{ij}^{m^n}}{r}\right) \tag{8}$$

In (8), d_{ij}^m denotes maximum difference (absolute) among the scalar components of our X_i^m and X_j^m, k is a constant whose value varies from 0 to 0.9 and r is k times SD (standard deviation) of the data, whereas m varies from 1 to 3 and n is gradient of the exponential function. Using D_{ij}^m, we constructed the function Φ^m as given in (9) and incremented m, (6) to (8) are repeated and Φ^{m+1} is obtained as in (10).

$$\varphi^m(n,r) = (N-m)^{-1} \sum\nolimits_{i=1}^{N-m} \left((N-m-1)^{-1} \sum\nolimits_{j=1, j\neq i}^{N-m} D_{ij}^m\right) \tag{9}$$

$$\varphi^{m+1}(n,r) = (N-m)^{-1} \sum\nolimits_{i=1}^{N-m} \left((N-m-1)^{-1} \sum\nolimits_{j=1, j\neq i}^{N-m} D_{ij}^{m+1}\right) \tag{10}$$

Finally, we defined FuzzyEn as in (11).

$$FuzzyEn(m, n, r, N) = \ln(\varphi^m(n, r)) - \ln(\varphi^{m+1}(n, r)) \tag{11}$$

Figure 4 shows a comparison in fuzzy entropy values for dataset S, F and Z. This FuzzyEn forms feature vector that are inputted to the SVM classifier for the classification of EEG data.

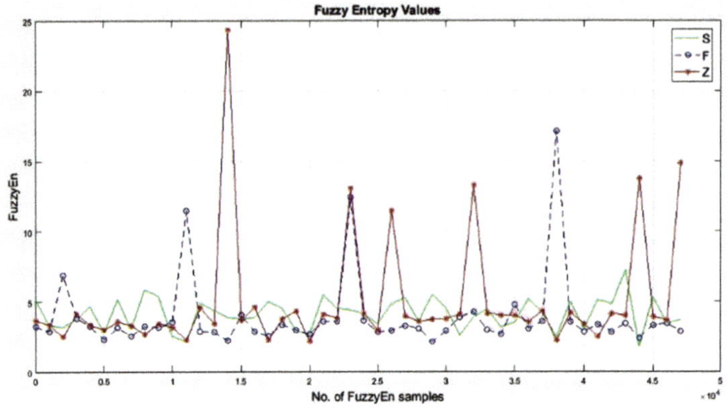

Fig. 4. FuzzyEn entropy plot of data set S, F and Z.

3.4 SVM Classifier

Due to its accuracy and capability to deal with large number of predictors, SVM is well known for binary classification problems in machine learning for high dimensional feature vectors. SVM attempts to design an optimal hyper-plane that maximizes the distance between this hyper-plane and the training points of both classes. Whether to use linear SVM or non-linear SVM is choice of concern. This work uses non-linear SVM with radial basis function (RBF) as kernel functions. For RBF, choice of sigma decides how the classifier behaves, that is for smaller values classifier gives strict decision boundaries whereas for larger values classifier considers nearly all the data points. In present work we used RBF as kernel function for SVM classifier with sigma as 0.2.

3.5 Statistical Parameters

The three parameters that are used in this work for performance assessment of SVM are sensitivity (SEN), specificity (SPE) and classification accuracy (CA) [7] and these are defined in (12), (13) and (14) respectively.

$$Sensitivity = \frac{true\,positives}{true\,positives + false\,negatives} * 100\% \tag{12}$$

$$Specificity = \frac{true\,negatives}{true\,negatives + false\,positives} * 100\% \tag{13}$$

$$Accuracy = \frac{correct\,classified\,patterns}{total\,patterns} * 100\% \tag{14}$$

4 Experimental Analysis

After extracting IMFs of the data, we divide the frame of 4097 samples into set of five different number of samples i.e. 173, 256, 512, 1024 and 2048 and FuzzyEn calculation is performed on these patterns.

Here during FuzzyEn calculation value of m varies from 1 to 3 yielding 3 values and that of k (our constant) varies from 0 to 0.9 in an interval of 0.1 yielding 10 values, thus there arises total 30 values per each time series. Table 1 gives final number of values obtained after calculation of all 100 text files of a given data set. It is clear that total number of samples before applying FuzzyEn were 4097 * 100 = 4, 09,700 which is now reduced.

Table 1. No. of FuzzyEn values for 100 time series.

Size of frame (N)	No of frame per each time-series	No of entropy values per each time-series	Total no. of entropy values
173	4097/173 = 23	(3 * 10) * 23 = 690	690 * 100 = 69000
256	4097/256 = 16	(3 * 10) * 16 = 480	480 * 100 = 48000
512	4097/512 = 8	(3 * 10) * 8 = 240	240 * 100 = 24000
1024	4097/1024 = 4	(3 * 10) * 4 = 120	120 * 100 = 24000
2048	4097/2048 = 2	(3 * 10) * 2 = 60	60 * 100 = 6000

This FuzzyEn forms input feature vectors that we apply to SVM by dividing it into training and testing dataset. Both datasets have a specific job as training dataset trains the SVM, whereas testing dataset verifies the accuracy of the trained SVM. Here initially entire data set is applied to the SVM for training and testing purpose. Specifically considering frame size of 1024 samples per frame, FuzzyEn generates 120 values per time series, leading to 12000 values for all 100 time series for one class (Z, O, F, N), similarly for comparison purpose 12000 values will also get generated for second class (S), resulting in total 24000 values. In the second phase of experiment, to decrease data dependency the training data set consist of 50% of total data and the remaining 50% is used for the testing of the classifier. In both cases performance evaluation is done using sensitivity, specificity and accuracy.

5 Experimental Results

In the present work feature extraction of the EEG data is done using EMD followed by FuzzyEn and SVM is used for the classification purpose. Here FuzzyEn is extracted with SD of the EEG data, and randomness of the data is extracted by having different values of m, r and N. Values of the statistical parameters obtained with 100% and 50% training of the classifiers are summarized in Table 2.

Table 2. Statistical parameters.

Training and testing	N	Cases for seizure detection	m	k	SEN (%)	SPE (%)	CA (%)
Without training	2048	Z-S	1	0	100	100	100
	2048	O-S	1	0	100	99.50	99.75
	1024	N-S	1	0	100	100	100
	2048	F-S	1	0	100	97.50	98.75
After 50% training	1024	Z-S	3	0	99.75	100	99.88
	1024	O-S	3	0.9	98.75	100	99.38
	1024	N-S	3	0	98.75	98.50	98.62
	2048	F-S	1	0.3	99	95	97

Figure 5 delivers classification results for all for cases with 50% training set graphically. Figure 6 delivers results for dataset Z and O respectively, when compared with S with 50% training.

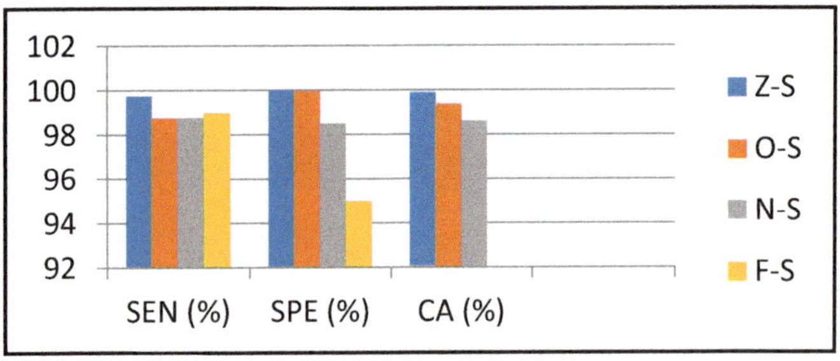

Fig. 5. Graphical representation of the statistical parameters with 50% training set.

6 Comparison with Other Works

Table 3 shows a comparison of the present work and previous works done in seizure detection process in terms of classification accuracy.

7 Conclusion

In the reported study, EEG signals were examined to constitute the classification feature vectors that were applied to SVM classifier for seizure detection. Results obtained here conclude that FuzzyEn can be considered as a better index when compared to mentioned methods.

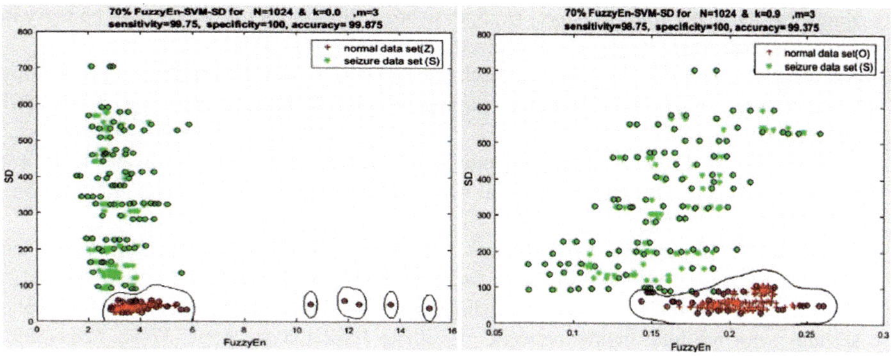

(a) **(b)**

Fig. 6. SVM classification using MATLAB software. (a) N = 1024, k = 0 and m = 3 for data set Z-S classification with SD. (b) N = 1024, k = 0.9 and m = 3 for data set O-S classification with SD.

Table 3. Comparison of CA of existing and proposed method.

Research group	Adopted index	Data set classified	CA (%)
Yuedong and Pietro [7]	SampEn	F-S	95.65
Guo et al. [8]	ApEn	Z-S	99.85
Nicolaou and Georgiou [9]	PE	Z-S	93.55
Kumar et al. [10]	WE	N-S	94.50
		Z-S	99.75
Guo et al. [16]	GP	Z-S	99.20
This study	FuzzyEn	Z-S	99.88
		O-S	99.38
		N-S	98.62
		F-S	97.00

References

1. Hao, Q., Gotman, J.: A patient-specific algorithm for the detection of seizure onset in long-term EEG monitoring: possible use as a warning Device. IEEE Trans. Biomed. Eng. **44**(2), 115–122 (1997)
2. Berger, H.: Uber das Elektrenkephalogramm des Menschen. Arch. f. Psychiat **87**(1), 527–570 (1929)
3. Guler, N., Vbeyli, E., Gular, I.: Recurrent neural networks employing Lyapunov exponents for EEG signal classification. Expert Syst. Appl. **29**(3), 506–514 (2005)
4. Lehnertz, K., Elger, C.: Spatio-temporal dynamics of the primary epileptogenic area in temporal lobe epilepsy characterized by neuronal complexity loss. Electroencephalogr. Clin. Neurophysiol. **95**(2), 108–117 (1995)
5. Accardo, A., Affinito, M., Carrozzi, M., Bouquet, F.: Use of the fractal dimension for the analysis of electroencephalographic time series. Biol. Cyber. **77**(5), 339–350 (1997)

6. Kannathal, N., Choo, M., Acharya, U., Sadasivan, P.: Entropies for detection of epilepsy in EEG. Comput. Methods Programs Biomed. **80**(3), 187–194 (2005)
7. Yuedong, S., Pietro, L.: A new approach for epileptic seizure detection: sample entropy based feature extraction and extreme learning machine. J. Biomed. Sci. Eng. **3**, 556–567 (2010)
8. Guo, L., Rivero, D., Pazos, A.: Epileptic seizure detection using multiwavelet transform based approximate entropy and artificial neural networks. J. Neurosci. Meth. **193**(1), 156–163 (2010)
9. Nicolaou, N., Georgiou, J.: Detection of epileptic electroencephalogram based on permutation entropy and support vector machines. Expert Syst. Appl. **39**(1), 202–209 (2012)
10. Kumar, S., Sriraam, N., Benakop, P., Jinaga, B.: Entropies based detection of epileptic seizures with artificial neural network classifiers. Expert Syst. Appl. **37**(4), 3284–3291 (2010)
11. Bruhn, J., Lehmann, L., Ropcke, H., Bouillon, T., Hoeft, A.: Shannon entropy applied to the measurement of the electroencephalographic effects of desflurane. Anesthesiology **95**(1), 30–35 (2001)
12. Chen, W., Zhuang, J., Yu, W., Wang, Z.: Measuring complexity using FuzzyEn, ApEn, and SampEn. Med. Eng. Phys. **31**(1), 61–68 (2008)
13. Cao, Z., Lin, C.: Inherent fuzzy entropy for the improvement of EEG complexity evaluation. IEEE Trans. Fuzzy Syst. **PP**(99), 1 (2017)
14. Andrzejak, R., Lehnertz, K., Rieke, C.: Indications of non-linear deterministic and finite dimensional structures in time series of brain electrical activity: dependence on recording region and brain state. Phys. Rev. E **64**, 1 (2001)
15. Ilyas, M., Saad, P., Ahmad, M.: Classification of EEG signals for brain-computer interface applications: performance comparison. In: 2016 International Conference on Robotics, Automation and Sciences (ICORAS), 5–6 November 2016, pp. 1–4 (2016)
16. Guo, L., Rivero, D., Dorado, J., Munteanu, C., Pazos, A.: Automatic feature extraction using genetic programming: an application to epileptic EEG classification. Expert Syst. Appl. **38** (8), 10425–10436 (2011)

Preliminary Study on Wearable Patch Device with EHG Envelope Peak as Feature Parameter for Preterm Birth Monitoring

Young Chang Jo(✉), Hyuck Ki Hong, Hae Na Kim, Won Hee Hwang, Yeon Shik Choi, and Suk Won Jung

Korea Electronics Technology Institute, SeongNam-Si, GyeongGi-Do 13509, South Korea
ycjo@keti.re.kr

Abstract. Uterine contraction is a critical activity of pregnancy monitoring and can be used to predict preterm birth. Recently, wearable uterine EMG measuring devices have been introduced to monitor pregnancy such as pre-term birth. In this work, we designed a prototype of abdomen attachable patch device as uterine EMG signal acquisition circuits. Also, the number of peak points of the EHG signal envelope was studied as new feature parameter for discrimination of preterm birth. Multi-channel uterine EMG signals could be measured simultaneously and the device size is about 60 * 50 mm^2. Multi-channel abdominal ECG+EMG signals of test subject could be measured successfully with this patch devices and the measured signals could be communicated wirelessly to other smart devices successfully. To extract EMG signal from ECG+EMG complex, we designed differential mode circuit configuration and test results shows good extraction performance. The number of envelope peak points of EHG signals as a new feature parameter is expected to be useful in predicting preterm delivery risk.

Keywords: Preterm birth · Monitoring · EMG · Patch

1 Introduction

Uterine activity assessment is essential for evaluating labor progress, and can be used to predict preterm labor. Currently widely used techniques for uterine contraction monitoring are invasive intrauterine pressure (IUP) measurements and external tocodynamometer method. While the first method is regarded as reliable, its invasive properties cause contamination and limit its application. External tocodynamometer is non-invasive, but is less reliable and often inconsistent due to the indirect mechanical nature of the strain gauge transducer used [1, 2]. An alternative is electrohysterography (EHG), the measurement of electrical signals resulting from the activity of uterine contraction EMG. EHG provides uterine activity information which is more accurate and reliable than tocodynamometer while maintaining the ease of use [3].

Recently, a few uterine EMG monitoring devices are introduced which measure uterine contraction activity such as Novii wireless patch of Monica Healthcare and Smart pregnancy tracker of Bloomlife. In this work, we introduced newly designed

S. O. Hwang et al. (Eds.): ICGHIT 2018, LNEE 502, pp. 119–124, 2019.
https://doi.org/10.1007/978-981-13-0311-1_21

multi-channel uterine EMG patch device prototype for preterm birth monitoring applications.

2 Patch Circuits Design

Technology concept of EHG (uterine EMG) based preterm birth monitoring is described in Fig. 1. The system consists of uterine EMG patch device and smartphone application. The measured EMG signals transmitted to smartphone via bluetooth telecommunication module. Technology concept of uterine EMG based pre-term birth monitoring. The proposed uterine EMG patch is designed as attachable type on abdomen of pregnant woman. Uterine EMG patch measure EHG (uterine EMG), FRH (fetal heart rate), maternal temperature and acceleration signals from physical activities of maternal abdomen. The simplified circuit block diagram of the designed uterine EMG patch devices is shown in Fig. 2. The main block of the circuits consists of uterine EMG (EHG) electrode, FHR (fetal heart rate) electrode, temperature sensor, 6-axis accelerometer IC, active filter circuits, signal amplifier circuits, 8 channel high precision high resolution ADC, MCU and Bluetooth telecommunication IC for mobile devices interface.

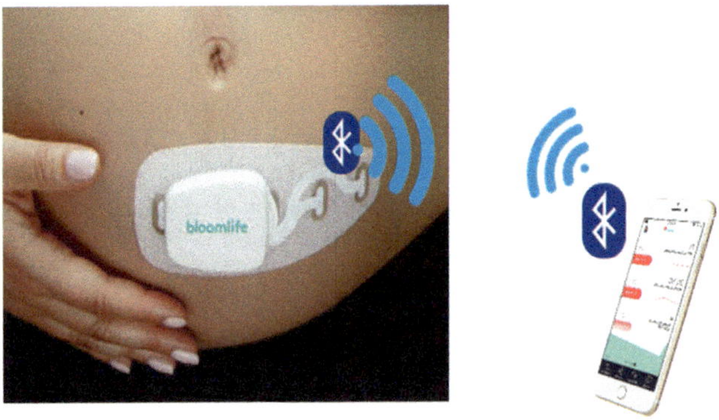

Fig. 1. Technology concept of uterine EMG based pre-term birth monitoring

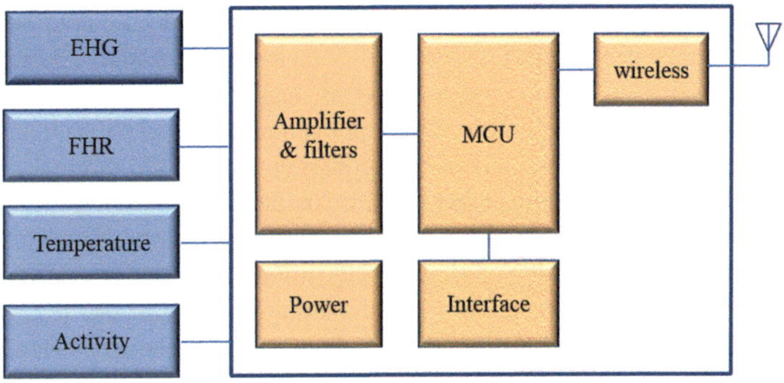

Fig. 2. Circuit block diagram of designed EHG (uterine EMG) patch devices

3 Implementation and Test Results

The proposed uterine activity monitoring patch and electrodes have been implemented in flexible PCB as shown in Fig. 3. The flexible PCB for electrodes are designed with shielded ground pattern for noise reduction. Figure 3 shows the prototype module of multi-channel uterine EMG patch with ADS 1299 IC. ADS 1299 IC is proper device for low amplitude bioelectrical signals such as ECG, EMG and EEG. ADS 1299 has 8 differential acquisition channels and each channel is equipped with a programmable gain amplifier, up to 24 times, and with an analog to digital delta-sigma converter of 24 bits resolution [4].

Fig. 3. The photograph of implemented uterine EMG patch devices with multi-channel electrodes

In the ADS1299 the oversampling technique is used to reduce the noise, spreading it on a wider band of 1.024 MHz. The only filtering achievable is a digital low pass decimation filter with the cut-off frequency that depends on the output data rate: lower data rate means narrower bandwidth and turns in higher SNR, higher data rate means wider bandwidth and thus more noise is included in the signals [4].

There are some complex bioelectrical signals in human abdominal area, such as ECG, EMG, etc. To extract genuine EMG signals we designed the circuits operated in differential mode which subtract one channel of signals from the other. Figure 4 shows the measured results of abdominal EMG+ECG signals during abdominal physical activity. In early stage of Fig. 4(a), there is only ECG signals because no abdominal physical activities are done in this stage. In next stage, there are ECG and EMG complex signals because some abdominal physical activities are done in this stage. Figure 4(a) shows measured EMG+ECG signals of channel 1 during abdominal physical activity.

Figure 4(b) shows EMG+ECG signals of channel 2 and Fig. 4(c) shows extracted EMG signals from difference between channel 1 and channel 2. Figure 5 shows the FFT plots of ECG and EMG respectively. EMG signals have higher amplitude than ECG signals and EMG signals have higher frequency components than ECG signals.

4 Feature Parameters

We could not obtain EHG (uterine EMG) data from pregnant women subjects because the proposed patch device has not yet been approved for pregnant women. Instead, we proceeded with feature parameter studies using existing EHG data such as Physionet. In this paper, the number of peak points of the EHG envelope was selected as a feature

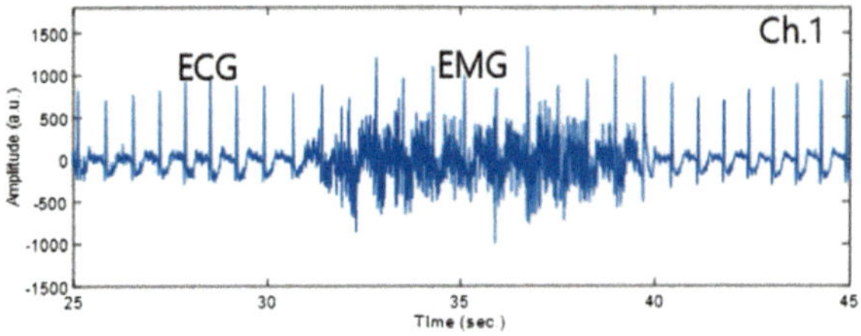

(a)

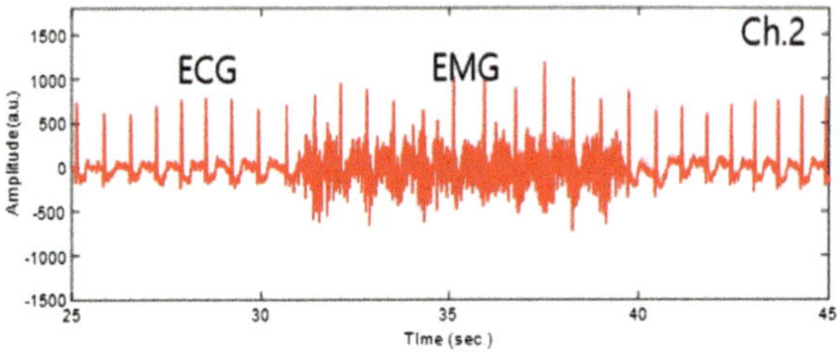

(b)

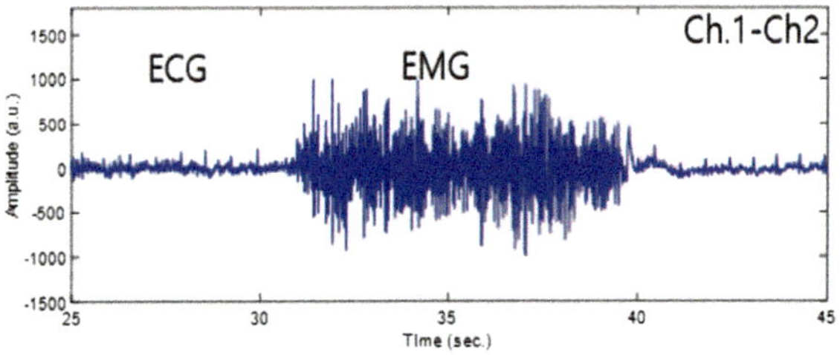

(c)

Fig. 4. The measured results of abdominal EMG+ECG signals during abdominal physical activity: (a) channel 1 signals, (b) channel 2 signals, (c) extracted EMG signals from difference between channel 1 and channel 2

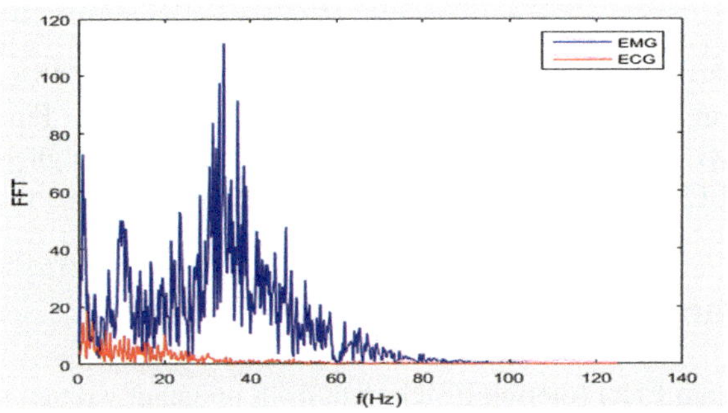

Fig. 5. The FFT plots of ECG and EMG respectively

parameter for discrimination of preterm birth. Envelope peak points feature extraction with cutoff frequency of 0.1 Hz for EHG data of 12 subjects selected from Physionet DB [5]. 6 data are for preterm birth and the other 6 data are normal case. The number of peak points of the EHG envelope was extracted and set as the threshold value twice the envelope mean. The uterine contraction characteristics were studied by calculating the envelope peak number larger than the set threshold.

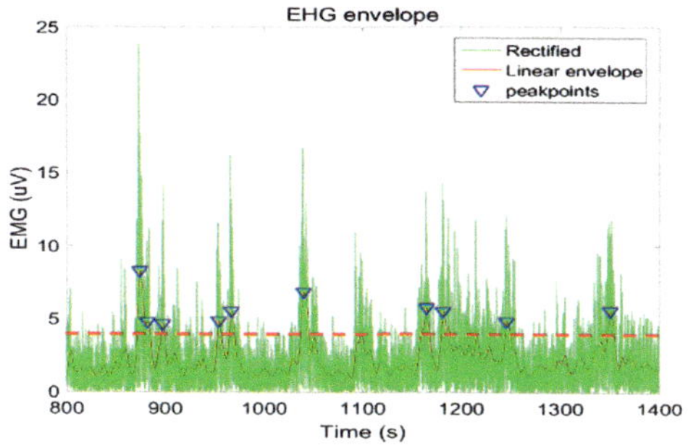

(a)

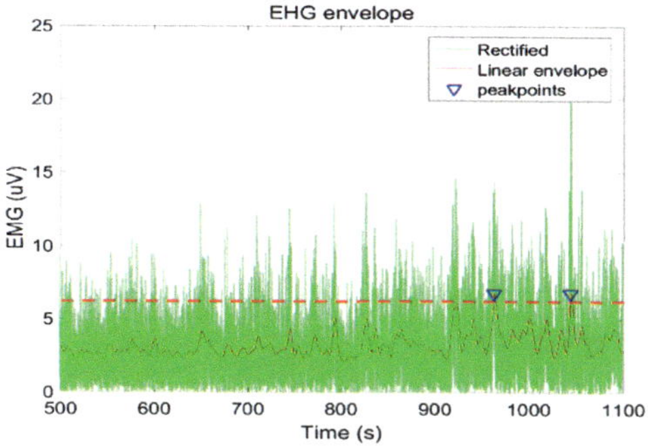

(b)

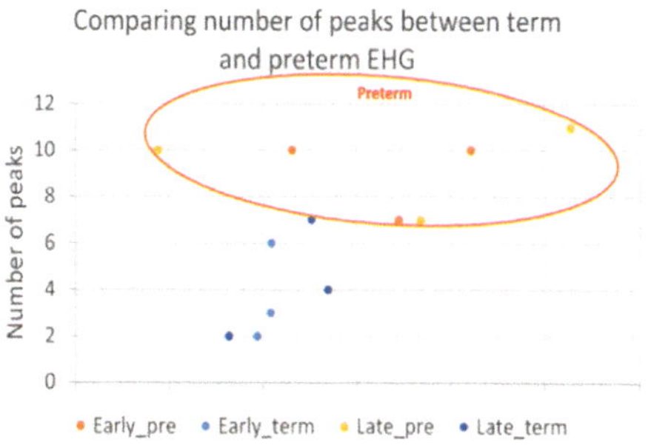

(c)

Fig. 6. The number of envelope peak points of EHG signals as new feature parameter for discrimination of preterm birth: (a) preterm birth, (b) normal pregnancy, (c) comparing plot between preterm birth and normal pregnancy

In the preterm birth case, more than 9 significant peak points were observed on average for 10 min as shown in Fig. 6(a). On the other hand, in the normal case, an average of 4 peak points were observed for 10 min as shown in Fig. 6(b). As shown in Fig. 6(c), the number of envelope peak points of EHG signals is expected to be useful in predicting preterm delivery risk.

5 Discussion and Conclusion

In this work, we designed a prototype of abdomen attachable patch device as EHG (uterine EMG) signal acquisition circuits to monitor pregnancy such as pre-term birth. Multi-channel EHG signals could be measured simultaneously and the device size is about 60 * 50 mm^2. Multi-channel abdominal ECG+EMG signals of test subject could be measured successfully with this patch devices and the measured signals could be communicated wirelessly to other smart devices successfully. To extract EMG signal from ECG+EMG complex, we designed differential mode circuit configuration and test results shows good extraction performance. The number of peak points of the EHG signal envelope was studied as new feature parameter for discrimination of preterm birth. In the preterm birth case, more than 9 significant peak points were observed on average for 10 min. On the other hand, in the normal case, an average of 4 peak points were observed for 10 min. The number of envelope peak points of EHG signals is expected to be useful in predicting preterm delivery risk.

Acknowledgement. This work was supported by the Korean Ministry of Science & ICT and IITP under project contract No. 2017-0-00741 [Development of intelligent pregnancy prediction system based on wearable patch type uterine EHG signal].

References

1. Bajlekov, G., Rabotti, C., Oei, S., Mischi, M.: Electrohysterographic detection of uterine contractions in term pregnancy. In: IEEE EMBC, Italy, pp. 5851–5854 (2015)
2. Felezorz, G., Kavsek, G., Antolic, N., Jager, F.: A comparison of various linear and non-linear signal processing techniques to separate uterine EMG record of term and pre-term delivery groups. Med. Biol. Eng. Comput. **46**(9), 911–922 (2008)
3. HayesGill, G., Hassan, S., Mirza, F., Cohen, W.: Accuracy and reliability of uterine contraction identification using abnormal surface electrodes. Clin. Med. Insights Women's Health **5**, 65–75 (2012)
4. Mastinu, B., OrtizCatalan, R., Håkansson, B.: Analog front-ends comparison: on the way to a portable, low-power and low-cost EMG controller based on pattern recognition. In: IEEE EMBC, Italy, pp. 2111–2114 (2015)
5. Physionet, The Term-Preterm EHG DB, Homepage. https://www.physionet.org/pn6/tpehgdb/. Accessed 2012

Study of Seizure Monitoring System Using Frontal Pole EEG Patch Device

Haena Kim, Won Hee Hwang, Hyuck Ki Hong, Yeon Shik Choi, Suk Won Jung, and Young Chang Jo(✉)

Korea Electronics Technology Institute (KETI), Gyeonggi, South Korea
ycjo@keti.re.kr

Abstract. For monitoring seizure signal from frontal pole EEG, normal EEG and ictal EEG of 10 subjects from free database were analyzed. Signal amplitude and prominent peaks of normal frontal pole EEG were compared to those of seizure EEG signal. As a result, seizure signals of all subjects had significant difference of prominent peak numbers compared to that of normal signals on the frontal location. We also fabricated a forehead-adherable patch for detecting frontal pole EEG signal.

Keywords: EEG patch · Epilepsy · Seizure · Monitoring · Frontal EEG Ictal EEG

1 Introduction

About 1% of the world's population is suffering from brain dysfunction, or epilepsy, caused by abnormal synchronous discharges of neuron aggregates in the brain [1]. Epilepsy is classified by causative factor of epilepsy. After the great discovery called Electroencephalogram (EEG) by Hans Berger in 1929, this biosignal has been used as an auxiliary method of detecting epilepsy [2]. The brain wave, EEG, electrophysiological signal occurred throughout the whole brain [3]. Brain wave is essential diagnostic signal for brain analyzing fields such as sleep quality or brain activity study. Recently, wearable device market has been expanded research for wearable-type EEG diagnostic tool is active. Especially in the case of epilepsy, patients are under the danger caused by unwarned seizures. Therefore, needs for easy and simple seizure detecting and warning system is on the rise. Lack of convenience and portability of conventional system detecting EEG is not appropriate for wearable device as the standard electrodes are positioned throughout the whole brain called 10–20 system or 10–10 system (Fig. 1).

In this paper, we designed 3 channels frontal EEG monitoring module for wearable device to overcome this shortcoming. Also, we confirmed the feasibility with patient EEG data from free database system if frontal pole signal has characteristics to detect seizure.

S. O. Hwang et al. (Eds.): ICGHIT 2018, LNEE 502, pp. 125–130, 2019.
https://doi.org/10.1007/978-981-13-0311-1_22

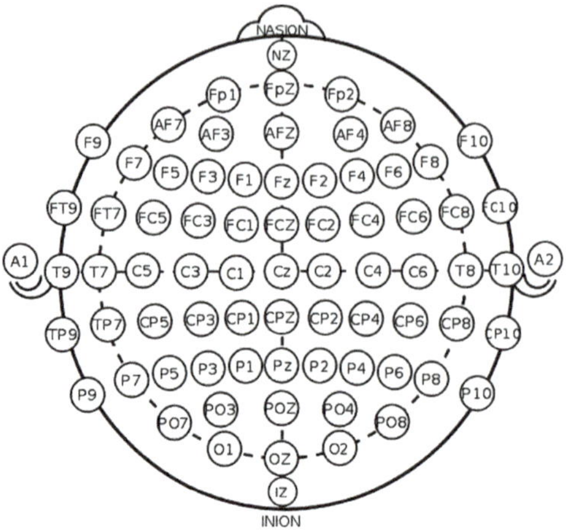

Fig. 1. Standard electrode positioning for detecting EEG

2 Experiments

2.1 Module Design

For measure the fine brain wave signals, we designed the circuit structure possible to process high – gain, high impedance and multiple channels. EEG amplification was adjustable from 1 to 24 times. Data transfer rate was designed to be controlled from 250 ksps to 16 ksps. RC circuit was also included for eliminating noise caused by impedance between skin and electrode. Electrode positions of fabricated module could be placed left (Fp1, F7, F3 in standard electrode positioning) or right (Fp2, F8, F4 in standard electrode positioning) frontal lobe (Fig. 2). We already confirmed this patch has the ability to measure the significant frontal EEG signals [4] when observing the brain activity differences.

Fig. 2. EEG patch prototype

2.2 Data Acquisition

To acquire the normal EEG signal and seizure signal from the same patient, we downloaded the data in physionet free database (physionet.org). The frontal EEG signal was extracted and analyzed in the same manner as the measurement with the fabricated module. A total of 10 subjects were selected regardless of gender, age and type of epilepsy (Table 1). Normal and ictal EEG have the same time length and normal EEG was extracted before 5 min from ictal EEG.

Table 1. Subjects information

ID	Gender	Age	ID	Gender	Age
A	F	11	**F**	F	1.5
B	M	11	**G**	F	14.5
C	F	14	**H**	M	3.5
D	M	22	**I**	F	10
E	F	7	**J**	M	3

3 Signal Processing

3.1 Amplitude Analysis

It is reported that certain ictal EEG shows high amplitude compared to the normal EEG. So the first analyzing step was amplitude comparison (Fig. 3). We tried to set the meaningful threshold level after observing amplitude difference. But half of subjects were well distinguished (A, B, C, G, I) but others not (F, D, E, H, J).

3.2 Prominent Peak Analysis

We extracted peaks with certain prominence from the EEG envelope calculated by coefficient frequency 5 Hz. Peak prominence was specified as 50,000 (Fig. 4). As shown above, there is significant difference between normal and ictal EEG peak prominence results. We counted prominent peak points higher than 50,000 (Fig. 5). As the results, the prominent peak numbers of ictal EEG is 6.3 times higher than that of normal on average. The patients who could not be detected with amplitude classification are easily distinguished by analyzing peak prominence (subject F, D, E, H, J). In the case of subject J, there are 1.5 times difference between normal and ictal EEG peak prominence result in spite of poor contact problem exists.

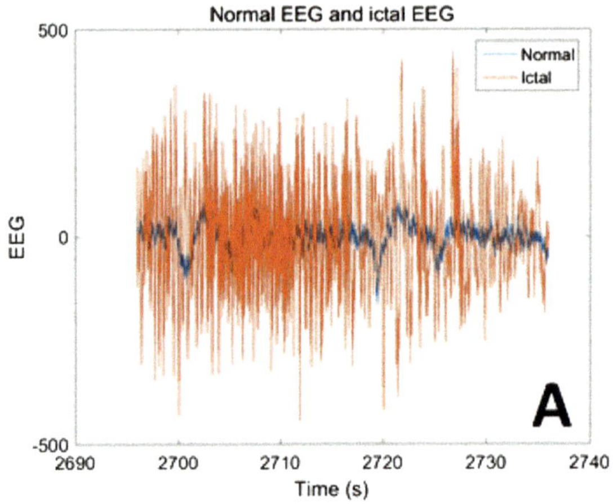

(a)

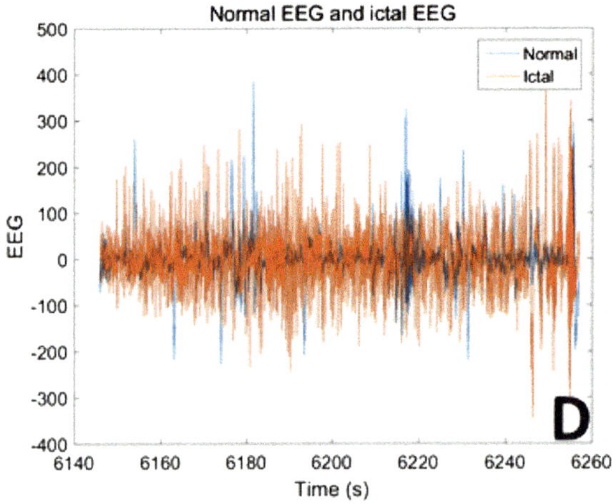

(b)

Fig. 3. Signal amplitude comparison between normal and ictal EEG. (a) EEG of subject A is clearly distinguishable but (b) ictal signal of subject D cannot be separated from normal signal.

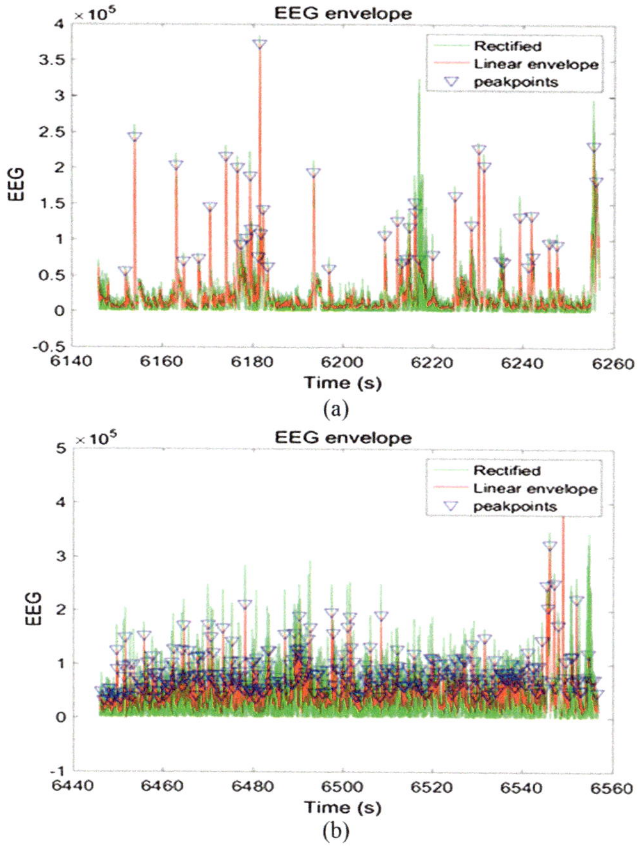

Fig. 4. Peak prominence analysis. Normal EEG (a) and ictal EEG (b) result of subject D. In the case of subject D, peak prominence analysis is more efficient than amplitude

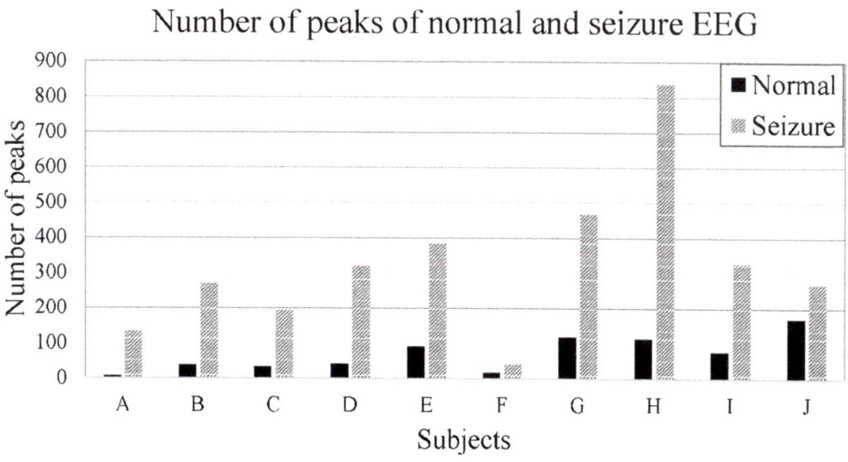

Fig. 5. Number of prominence peaks' difference between normal and ictal EEG

4 Conclusion

For patients suffering from the unwarned epileptic seizures, we designed easy and simple frontal EEG seizure detection patch attachable to the forehead. Using physionet free EEG database source, the feasibility of detecting seizure signals from frontal EEG is confirmed. From the EEG envelope, we extracted peak points with certain prominence higher than 50,000. The numbers of extracted peak points are different 6.2 times on average between normal and ictal EEG. 3 channel frontal EEG patch fabricated in this paper will be expected as a simple seizure warning system that can detect seizures by attaching to the forehead.

Acknowledgement. This work was supported by the Korean Ministry of Trade, Industry & Energy and KEIT under project contract No. 10053091 [Development of human-body-friend smart patch and home-based healthcare service solution for monitoring acute/chronic brain/cardiovascular diseases].

References

1. Mormann, F., Andrzejak, R.G., Elger, C.E., Lenhnertz, K.: Seizure prediction: the long and the winding road. Brain **130**(2), 314–333 (2007)
2. Kim, S.: Classification of epileptic seizure. J. Kor. Epil. Soc. **3**(2), 213–222 (1999)
3. Jin, B.: Clinical Physiology EEG. Korean medical book publishing company, Seoul (2014)
4. Jo, Y., Kim, H., Hwang, W., Hong, H., Choi, Y., Jung, S.: A single channel frontal pole EEG patch devices for sleep monitoring applications. In: CONFERENCE 2017, KICS, vol. 1, pp. 631–632 (2017)

Association Analysis of Postoperative Hypothermia for Stability of Blood Pressure

Hoon Jin[1] and Yong-Gyu Jung[2(✉)]

[1] Kyonggi University, 154-42, Gwanggyosan-ro, Yeongtong-gu, Suwon-si, Gyeonggi-do, Republic of Korea
bioagent@kyonggi.ac.kr

[2] Eulji University, 553, Sanseong-daero, Sujeong-gu, Seongnam-si, Gyeonggi-do, Republic of Korea
ygjung@eulji.ac.kr

Abstract. Hypothermia is defined as a body temperature (core, or internal body temperature) of less than about 95 F (35 C). Usually, hypothermia occurs when the body's temperature regulation is overwhelmed by a cold environment. However, in the medical and lay literature there are essentially two major classifications, accidental hypothermia and intentional hypothermia. Intentional hypothermia is body temperature lowering induced usually for a medical procedure. In the study, it has been shown that using k-means methods and association rules with post-operative patient data from a medical point of view. The results of the study are derived. In addition, two methods are used to derive the results related to stability of blood pressure, present and analyze them, and suggest future research directions.

Keywords: Hypothermia · Blood pressure stability · K-means
Association rules · Post-operative patient

1 Introduction

Infinite amount of information is leaked and flooded, which is overlooked in modern society. But information is very valuable to understand. For patients who have been treated are prolonged and interest in health is high and finding measures to prevent disease recurrence is very valuable. The data for the post-operative patient in the medical field is suggesting the direction of the patient after surgery because of the concern about "hypothermia". Hypothermia is defined as a body temperature (core, or internal body temperature) of less than about 95 F (35 C). Usually, hypothermia occurs when the body's temperature regulation is overwhelmed by a cold environment. However, in the medical and lay literature there are essentially two major classifications, accidental hypothermia and intentional hypothermia. Accidental hypothermia usually occurs from an exposure to cold that results in lowering the body temperature. Intentional hypothermia is body temperature lowering induced usually for a medical procedure. Although it is important that the disease does not occur, the issue is selected because the possibility of recurrence can not be overlooked. This data will specifically focus on blood pressure stability of post hypothermic patients. The relationship

S. O. Hwang et al. (Eds.): ICGHIT 2018, LNEE 502, pp. 131–136, 2019.
https://doi.org/10.1007/978-981-13-0311-1_23

between the stability of the patient's internal and surface temperature and blood pressure stability and blood pressure stability according to temperature and blood pressure measurement will be discussed. In addition, we divide the results into the clusters, and then we divide the output of internal temperature, surface temperature, final blood pressure measurement, and blood pressure stability into groups.

Hypothermia is a potentially dangerous drop in body temperature, usually caused by prolonged exposure to cold temperatures. The risk of cold exposure increases as the winter months arrive. But if you're exposed to cold temperatures on a spring hike or capsized on a summer sail, you can also be at risk of hypothermia. When the balance between the body's heat production and heat loss tips toward heat loss for a prolonged period, hypothermia can occur. Accidental hypothermia usually happens after cold temperature exposure without enough warm, dry clothing for protection. Mountain climbers on Mount Everest avoid hypothermia by wearing specialized, high-tech gear designed for that windy, icy environment. However, much milder environments can also lead to hypothermia, depending on a person's age, body mass, body fat, overall health, and length of time exposed to cold temperatures. The purpose of this paper is to compare the two methods from the experimental results and draw them to the conclusion, which is helpful for post - patient management.

2 Related Algorithms

2.1 K-Means Algorithm

In the k-Means algorithm, the labeling function is computed by comparing the distances of a data point xi from the vectors which represent the clusters (the centroids cj). The centroids are the model parameters which are estimate by using iterative steps. The k-means algorithm defines the centroid of a cluster as the mean value of the points within the cluster.

Algorithm:

The k-means algorithm for partitioning, where each cluster's center is represented by the mean value of the objects in the cluster.

Input: c: the number of clusters and D: a dataset containing n objects

Output: A set of k clusters

Method:

i. Arbitrarily chose k objects from D as the initial cluster centers;

ii. Repeat

iii. (re)assign each object to the cluster to which the object is most similar, based on the mean value of the objects in the cluster;

iv. update the cluster means, that is calculate the mean value of the objects for each cluster;

v. until no change

The time complexity of the k-means algorithm is O(nkt), where n is the total number of objects, k is the number of clusters, and t is the number of iterations. That is, for fixed k and t, the time complexity is O(n). Therefore, the method is relatively scalable and efficient in processing large datasets.

2.2 Association Rule

Association rule learning is a rule-based machine learning method for discovering interesting relations between variables in large databases. It is intended to identify strong rules discovered in databases using some measures of interestingness. Association rule mining is a procedure which is meant to find frequent patterns, correlations, associations, or causal structures from data sets found in various kinds of databases such as relational databases, transactional databases, and other forms of data repositories. Given a set of transactions, association rule mining aims to find the rules which enable us to predict the occurrence of a specific item based on the occurrences of the other items in the transaction. Association rule mining is the data mining process of finding the rules that may govern associations and causal objects between sets of items.

Association rule mining finds interesting associations and correlation relationships among large sets of data items. Association rules show attribute value conditions that occur frequently together in a given data set. The first number is called the support for the rule. The support is simply the number of transactions that include all items in the antecedent and consequent parts of the rule. The support is sometimes expressed as a percentage of the total number of records in the database. The other number is known as the confidence of the rule. Confidence is the ratio of the number of transactions that include all items in the consequent, as well as the antecedent (the support) to the number of transactions that include all items in the antecedent.

Lift is one more parameter of interest in the association analysis. Lift is nothing but the ratio of Confidence to Expected Confidence. Using the above example, expected Confidence in this case means, "confidence, if buying A and B does not enhance the probability of buying C." A lift ratio larger than 1.0 implies that the relationship between the antecedent and the consequent is more significant than would be expected if the two sets were independent. The larger the lift ratio, the more significant the association.

3 Experiments

Experimental data are on hypothermia, which can become an important problem for patients after surgery. Various data such as surface temperature, internal temperature, last blood pressure measurement, and oxygen saturation are presented, and the result is set to be expressed as blood pressure stability. All properties have nominal characteristics. The total number of data is 90 and the characteristics are classified into 9 kinds (Table 1).

Experiments is based on post-operative patient data and set the subject to be BP-STBL (blood pressure stability) as a subject related to blood pressure. First, in the k-means method, the input values are divided into five clusters and classified into L-CORE, L-SURF, and L-BP. The following figure shows the cluster size.

Table 1. Properties of experimental data

L-CORE	Patient's internal temperature in C high(>32), mid(>=36 and <=37), low(<36)
L-SURF	patient's surface temperature in C
L-02	oxygen saturation in % excellent(>=98), good(>=90 and <98), faire(>=80 and <90), poor(<80)
L-BP	last measurement of blood pressure high(>130/90), mid(<=130/90 and >=90/70), low(90/70)
SURF-STBL	stability of patient's surface temperature stable/mod-stable/unstable
CORF-STBL	stability of patient's core temperature stable/mod-stable/unstable
BP-STBL	stability of patient's blood pressure stable/mod-stable/unstable
COMFORT	patient's perceived comfort at discharge, measured as an integer between 0 and 20
decision ADM-DECS	I (patient sent to intensive care unit) S (patient prepared to go home) A (patient sent to general hospital floor)

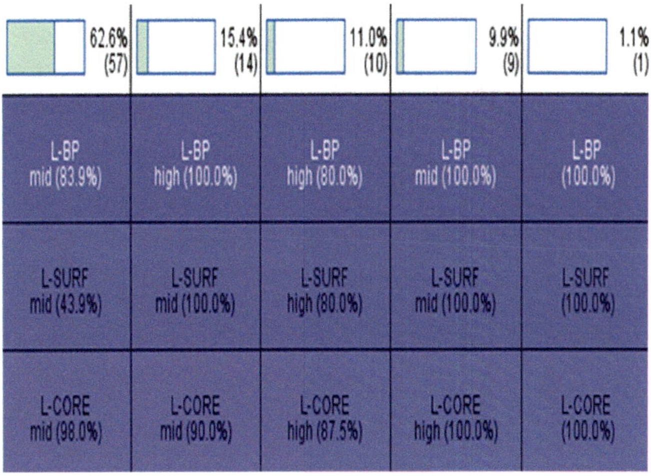

Fig. 1. Experimental results of k-means clustering

Next, the method of correlation analysis is as follows. The value of the previous term is set to L-CORE, L-SURF and L-BP, and the last value is set to BP-STBL. And reliability. It is possible to grasp the relation between the attributes with high reliability. It shows the relationship between the stability in the association analysis. The previous values are set to CORE-STBL and SURF-STBL, and the following values are set to the

BP-STBL = stable	L-CORE = low L-SURF = mid L-BP = mid	80.0
BP-STBL = stable	L-CORE = low L-BP = mid	76.923
BP-STBL = stable	L-BP = high L-SURF = mid L-CORE = mid	75.0
BP-STBL = stable	L-CORE = low L-SURF = low	71.429
BP-STBL = stable	L-SURF = mid L-CORE = mid	67.647
BP-STBL = unstable	L-CORE = high L-BP = high	66.667
BP-STBL = stable	L-CORE = low L-SURF = low L-BP = mid	66.667
BP-STBL = stable	L-SURF = mid L-BP = mid L-CORE = mid	64.0
BP-STBL = stable	L-SURF = mid L-BP = mid	63.636
BP-STBL = stable	L-CORE = low	63.158
BP-STBL = stable	L-SURF = mid	60.417

Fig. 2. Experimental results of correlation analysis

default value BP-STBL. This was also listed in the order of highest reliability and the following results were obtained.

Sample Heading (Forth Level). The contribution should contain no more than four levels of headings. The following **오류! 참조 원본을 찾을 수 없습니다.** gives a summary of all heading levels.

4 Conclusion

This paper focuses on medical and data mining, which is inseparable from contemporary society, and merges the two to present research and its results. We concluded that blood pressure stability is low for patients with high internal temperature based on the reliability of blood pressure stability for patients with hypotensive surgery and concluded that there is no correlation between internal temperature and surface temperature. We used the method of data mining research of k-means and association law. Other data related to blood pressure stability, including internal temperature, surface temperature, and final blood pressure measurement, will be used to provide information on blood pressure stability and its results. As there are various analysis methods in data mining, we will study different methods of analysis in a more accurate and helpful way than existing papers.

Previously, the target was set as blood pressure stability, and the relationship and clustering between various properties were achieved. First, among the five clusters in the k-means, L-BP has the highest mid-value and L-SURF and L-CORE have the highest mid-value. From the results, it can be seen that as the size of the cluster decreases, mid, which means normal, decreases in all areas of L-BP, L-SURF and L-CORE. In the case of the correlation analysis, the BP-STBL did not have any relation to L-SURF and L-BP in the case of Fig. 1, but the results were related to L-CORE (internal temperature). If the internal temperature is low or normal, it is stable, but if the internal temperature is higher than 37 °C, the blood pressure becomes unstable. On the other hand, referring to Fig. 2, internal temperature stability, surface temperature stability the blood pressure stability is not remarkably correlated. The results were derived using two sets of k-means and associativity rules, and the results were designed for blood pressure stability.

Acknowledgement. This research was supported by Next-Generation Information Computing Development Program through the National Research Foundation of Korea (NRF) funded by the Ministry of Science and ICT (2017M3C4A7083412).

References

1. Moore, W.C., et al.: Identification of asthma phenotypes using cluster analysis in the severe asthma research program. Am. J. Respir. Crit. Care Med. **181**(4), 315–323 (2010)
2. Tarassenko, L., Hann, A., Young, D.: Integrated monitoring and analysis for early warning of patient deterioration. Br. J. Anaesth. **97**(1), 64–68 (2006)
3. Tomar, D., Agarwal, S.: A survey on data mining approaches for healthcare. Int. J. Bio-Sci. Bio-Technol. **5**(5), 241–266 (2013)
4. Vogt, W., Nagel, D.: Cluster analysis in diagnosis. Clin. Chem. **38**(2), 182–198 (1992)
5. Ian, H.W., Eibe, F., Mark, A.H.: Data Mining Practical Machine Learning Tools and Techniques, 3rd edn. Morgan Kaufmann Publishers (2011)

Analysis of the Protection Circuit in the Function Generator for Education

SungYeol Kwon[1], HyunChang Lee[2(✉)], KyuTae Lee[2], HyunMook Cho[2], and SangYep Nam[3]

[1] PuKyong National University, PuSan 48513, Korea
sungyeol@pknu.ac.kr
[2] KongJu National University, CheonAn 31080, Korea
{hclee,ktlee,hmchov}@kongju.ac.kr
[3] KookJe University, PyeongTaek 17731, Korea
synam58@gmail.com

Abstract. In this paper, when the function generator that essential equipment in the electronic engineering is used for educational purposes, the cause of the failure was analyzed and methods for preventing it were proposed. The proposed method can significantly reduce equipment damage. Therefore, education can proceed more efficiently in terms of time and cost.

Keywords: Function generator · Protection circuit · Booster · Current-limiter

1 Introduction

Function generator is one of the essential equipment which must be used by educational institutions, laboratory and business. Users who are familiar with the function generator don't matter for using it, but users who are not familiar with it could break it by mistakes.

Failure of function generator can result in financial loss as well as disruption of lesson due to a lack of available equipment. In the worst case, it can cause accident such as fire. Therefore the equipment to be used for education requires a protection circuit. In this paper, the failure cases of function generators used in educational institutions are analyzed and a method is proposed to prevent them.

2 Case of Failure in Function Generator

2.1 Mechanical Failure

In general, the frequency of the function generator is controlled using a 10-turn resistor configured as a dial [1]. A dial-panel part which consists of the toothed wheels for display the status of 10-turn is broken due to the excessive force. This failure is caused by the low cost of mechanical parts. Therefore, even if the price rises, it is necessary to use more robust parts for educational purposes.

S. O. Hwang et al. (Eds.): ICGHIT 2018, LNEE 502, pp. 137–140, 2019.
https://doi.org/10.1007/978-981-13-0311-1_24

2.2 Output Booster Amplifier

In the education field, the most damaged component in the function generator is the output booster amplifier. Because the trainee is not familiar with electronic equipment, short to ground or connect to supply can damage the components of the function generator such as boost transistor, swamping resistor and oscillator. Moreover, since the most parts of function generator are assembled using SMD (Surface Mount Device) type, it is impossible for the educator to find the fault and repair it.

3 Protection Methods for Function Generator

3.1 Protection for Connect to Supply

Figure 1 show a general output booster used in a function generator. Push-pull amplifiers use two complementary transistors (Q_1 and Q_2) with swamping resistor (R_3 and R_4). Diode D_1 and D_2 are used in order to eliminate crossover distortion [2]. And R_5 is the output protection resistor.

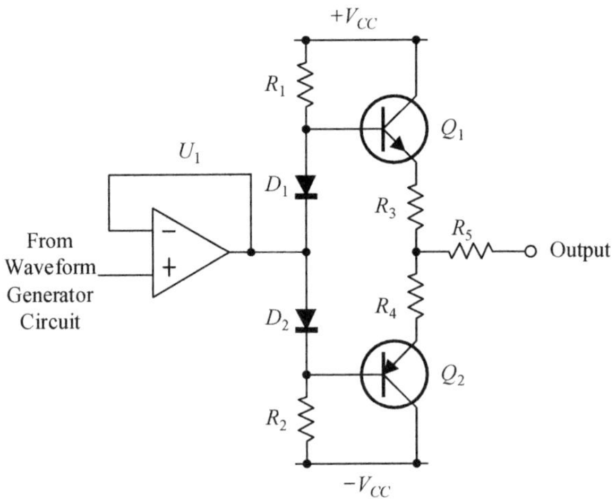

Fig. 1. The conventional output booster amplifier circuit

In this circuit, if the output is short to ground and output is not maximum voltage, R_5 protects the circuit by limiting the current. However, if the maximum output lasts for a long time, R_5 is broken, and Q_1, Q_2, and swamping resistors R_3 and R_4 are also broken. Setting the protection resistor R_5 to a large can protect the circuit, but the output impedance of the circuit is increased and the load current is limited [3]. Generally, the function generator is configured to select two resistors such as 47Ω and 150Ω, which is not useless to the trainee who is not used to the electronic device. Therefore it is necessary to add a current limit circuit to limit the current supplied by the output transistor as shown in Fig. 2. Since the current in the circuit is limited, the current limit resistor R_5 connected to the output can be removed. The limited current is obtained by (1).

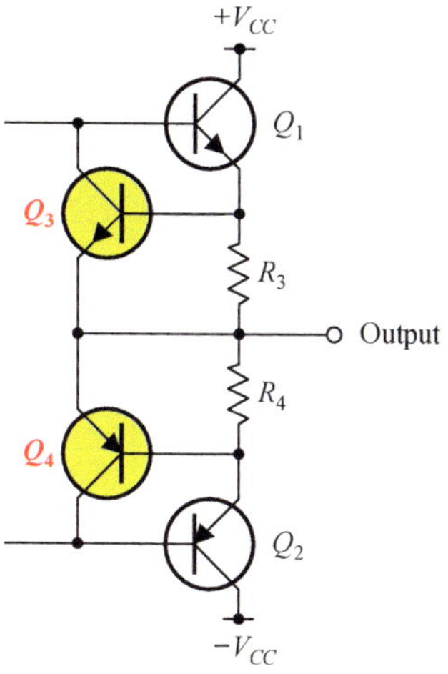

Fig. 2. The output booster with current limit circuit

$$I_{\max} = \frac{V_{BE3}}{R_3} = \frac{V_{BE4}}{R_4} \tag{1}$$

3.2 Reverse Voltage Protection

In Fig. 2, when the output is connected to the positive supply voltage, current flows through R_4, Q_2 and Q_2 can be protected by the current limit circuit. However, when the function generator output goes negative voltage, reverse voltage is applied to Q_1. In general, since the base-emitter reverse voltage of a transistor is very small [4], Q_1 is damaged. When the output is connected to the negative supply voltage, Q_2 stage is damaged for the same reason. Therefore, the reverse voltage protection is essential as shown in Fig. 3. In the Fig. 3, diode D_3 and D_4 are reverse voltage protection, and D_5 and D_6 are diodes for bias compensation according to addition of D_3 and D_4.

3.3 Distortion of Output Waveform

As previously mentioned, if a current limit circuit and a reverse voltage protection circuit are added to the output stage, there is a difference between the U_1 output and the amplifier output. Therefore the buffer needs to sense the final output to compensate for distortion of the waveform as shown in Fig. 3.

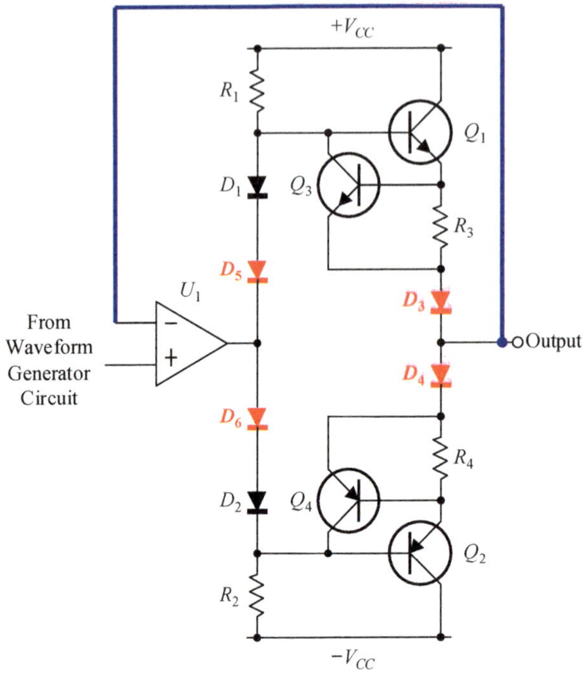

Fig. 3. The output booster with reverse voltage protection

4 Conclusion

In this paper, when the function generator that essential equipment in the electronic engineering is used for educational purposes, the cause of the failure was analyzed and methods for preventing it were proposed. The proposed method can significantly reduce equipment damage. Therefore, education can proceed more efficiently in terms of time and cost.

References

1. BOURNS: Precision Potentiometers Multiturn. http://www.bourns.com/products/potentiometers/precision-pots-multiturn. Accessed 5 January 2018
2. Malvino, A., Bates, D.J.: Electronic Principles, 7th edn. McGraw-Hill, New York (2006)
3. Neudeck, G.W., Hayt, W.H.: Electronic Circuit Analysis and Design, 2nd edn. Pearson Education, Upper Saddle River (1984)
4. ON Semiconductor: Understanding Power Transistors Breakdown Parameters - Rev. 3. http://www.onsemi.com/pub/Collateral/AN1628-D.pdf. Accessed 5 January 2018

Development of DC Smart Plug Control System

Narakorn Songkittirote[1], Worajit Setthapun[1], Kobsak Sriprapha[2], and Hathaithip Ninsonti[1](✉)

[1] Asian Development College for Community Economy and Technology, Chiang Mai Rajabhat University, Chiang Mai, Thailand
hathaithip.nin@gmail.com

[2] National Electronics and Computer Technology Center, Pathum Thani 12120, Thailand

Abstract. This work focused on the development of DC smart plug for efficient data monitoring and transmission. The transmitted data were collected from smart plug sensors and user commands, and sent via wireless transmission to the online database. The results revealed that the smart plug could control and provide high speed wireless data transmission with minimal error to and from the appliances. The online media was also developed with the smart plug display the power usage and statistics.

Keywords: DC smart plug · Data monitoring · Cloud server Electrical appliances control

1 Introduction

Controlling electrical appliances is a very important topic because this will assist in changing the energy usage behavior. In order to control the electrical appliances, an efficient control system must be developed [1, 2]. In this work, the smart plug was developed as the fundamental controlling device for the electrical appliances. A plug is the key connection between the electrical source and the appliances. With the development of a smart plug, it will be able to control several types of appliances. This research focused on the development of DC Smart Plug to facilitate the energy efficiency of the appliances [3]. The usage of Direct Current (DC) in the appliances would minimize the loss from current conversion from AC to DC [4]. The smart plug could control the appliance with accurate and efficient wireless data transmission. The communication algorithm between the controlling devices and the online system would provide the decision factors for automatic control or by user setting. In addition, power usage data could be collected to the online database and SD card.

Various researchers also focused on the appliance controlling technologies. Tang presented a model for Smart Home light controls [5]. Risteska reported on the challenges and solutions for Internet of Things of Smart Home [6]. Both groups provided the advantages, disadvantages, and concepts of wireless control systems for electrical appliances. The advantages of wireless system were the convenient, less complication from wiring, and low cost when compared to the wired control system. In addition, this

S. O. Hwang et al. (Eds.): ICGHIT 2018, LNEE 502, pp. 141–148, 2019.
https://doi.org/10.1007/978-981-13-0311-1_25

type of wireless system could be implemented to other application such as controlling water distribution wirelessly in Smart Farm [7]. The disadvantage of wireless system was the quality of wireless signal referred in the work of Leonardi. The home environment area affected the quality of wireless signal [8]. One of the method for wireless control were using WiFi chipset module. For example, Kazala focused on the Wireless Network for Mobile Robot Applications [9], Silveira studied Temperature Monitoring Through Wireless Sensor Network [10] and Kelly worked on Ambient and laboratory evaluation of a low-cost particulate matter sensor [11]. These 3 research groups considered chipset ESP8266 that is small, affordable and efficient. Programming could also be developed to fix the flaws of the wireless signal. In addition, MQTT Protocol could be applied to that chipset to enhance the quality of the wireless signal. MQTT could transmit high speed data in low quality signal. For example, Komkrit and coworkers revealed that MQTT could be used for Heart Rate monitoring [12]. Heart rate monitoring required the stable wireless signal and MQTT exhibited high efficient application.

Therefore, the appliance control system relied on the software to enhance the efficiency of the hardware system to minimize the flaws and instabilities of the wireless system. Controlling appliances through online service would reduce the working process and provide an efficient way for wireless data transfer for the smart plug.

2 Monitoring and Data Transmission for DC Smart Plug

2.1 Controlling Device

Figure 1 shows the major components of the DC Smart Plug control system. Microcontroller unit (MCU) in this work used ESP826. This WiFi chipset module is small, energy efficient, and versatile for the applicant of Client, Access Point and Client+AP [13]. The chipset would receive commands or data from the online system for controlling the smart plug. During the controlling process, the MCU would receive the command from

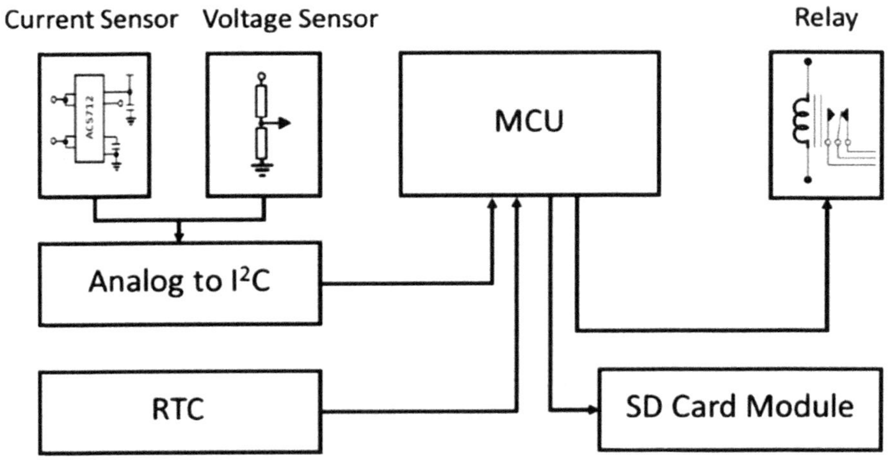

Fig. 1. Major components of a control system

sensors that measured the energy usage of the smart plug. Then, the energy usage data would be processed to control the ON/OFF mode of the appliances. Relay would be used as the electromagnetic switch for the electrical circuit control [14]. In addition to controller the appliances, the smart plug would save the energy usage data for each time period onto the SD Card for further control analysis.

2.2 Power Consumption Display System

Voltage and current sensors were installed in order to calculate power consumption of each electric appliance connected to DC smart plug. Figure 2 shows diagram of sensors and microcontroller unit (MCU). ACS712 current sensor chipset based on the principle of Hall-effect were used. This sensor has a current range of 0–5 A [15]. For voltage sensor, voltage divider was used to produce an output voltage of 0–5 VDC from an input voltage of 0–300 VDC [16]. Output data from current and voltage sensors were converted to digital number by chipset. Then, it was transfer to MCU with I2C Protocol. It could support multiple sensors connected to the smart plug [17].

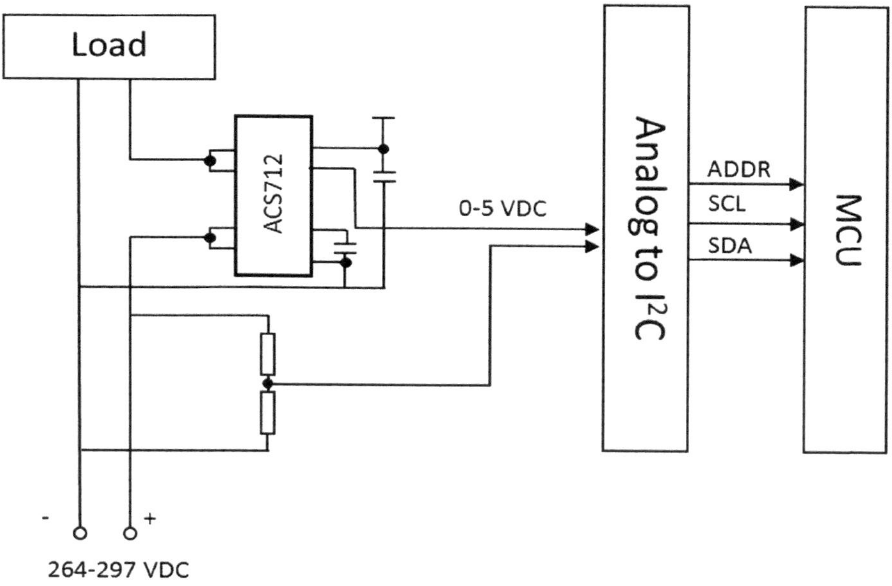

Fig. 2. Diagram of sensors and *microcontroller* unit

2.3 Online Control

The online control system produced smaller Hardware (ESP8266) and lower memory usage (1 kB RAM) for the smart plug [18]. The Cloud Server provided faster processing and smaller commands prior to transmitting to the smart plug (Fig. 3). The Cloud Server was responsible for receiving the user commands through various devices. Web Service was used to process commands and exchange data through the network [19]. The Web Service was designed to increase the speed of smart plug

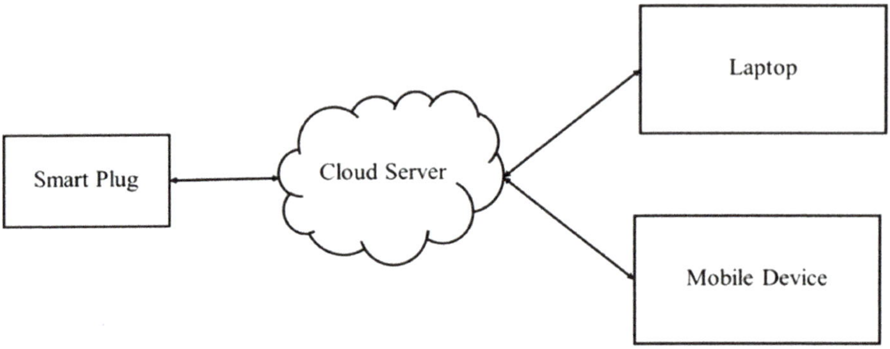

Fig. 3. Data transfer via Cloud Server

control via the internet. In addition, the hardware design also enhanced the speed of the data transfer.

2.4 Smart Plug Control via Application

Figure 4 exhibits the example of the Online Application to control the Smart Plug. The control system comprised of 5 functions. (1) Login function for system security;

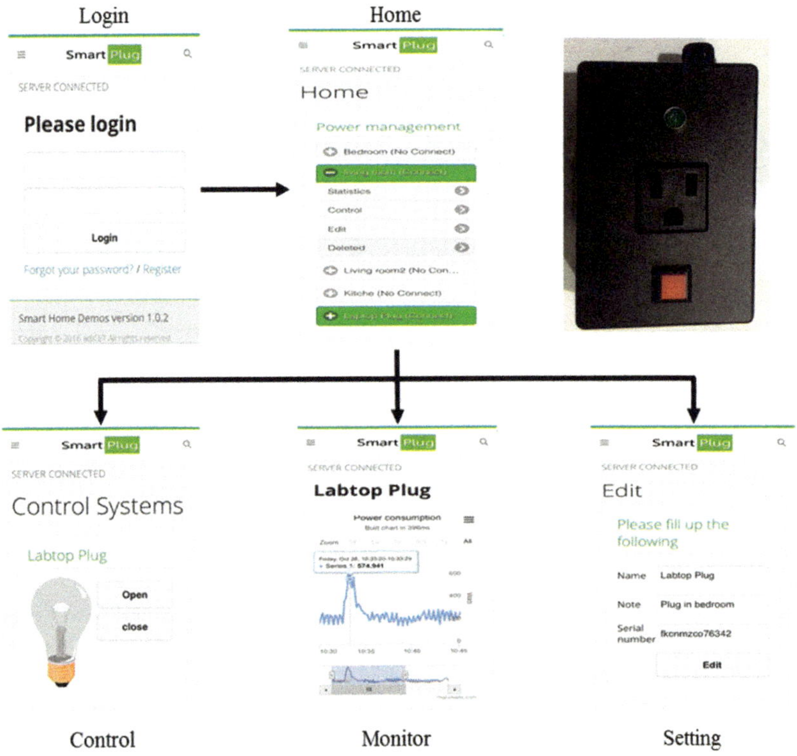

Fig. 4. Example of mobile application

(2) Status function for the connection status of the plug and status of appliance control; (3) Manual On/Off function and display; (4) Power consumption display function - graphical statistics display format; and (5) Smart Plug problem solving function.

3 Results and Discussion

3.1 Voltage and Current Sensors Evaluation

In this research, power consumption was determined using voltage and current sensors. The sensors were evaluated as shown in Figs. 5 and 6. The results showed that voltage sensor and current sensor have an error of 2.02% and 2.70%, respectively.

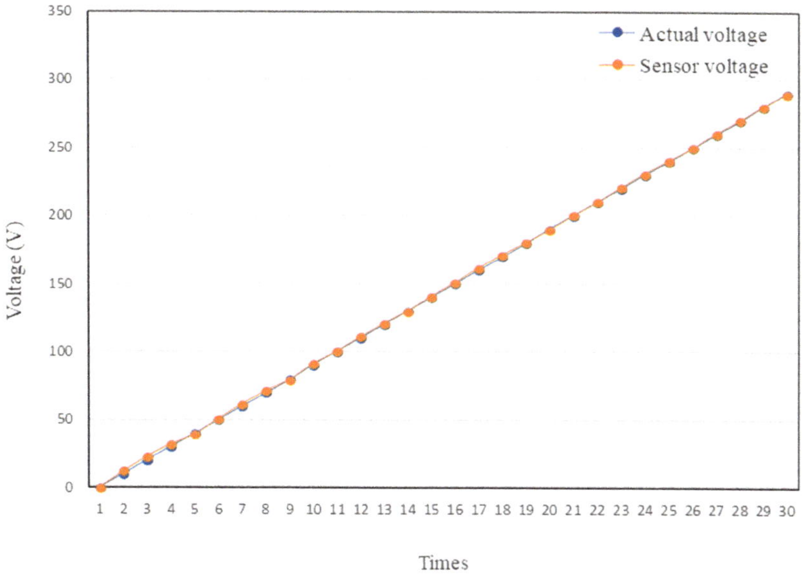

Fig. 5. Voltage sensor evaluation

3.2 Data Transfer Evaluation

The process of sending and receiving power consumption data via wireless networks was tested as shown in Fig. 7. It was found that no error occurred from 100 times of testing. From the observation, the duration of each data transfer is within 0.8 min. However, there are about 4% outliers of the average transfer time.

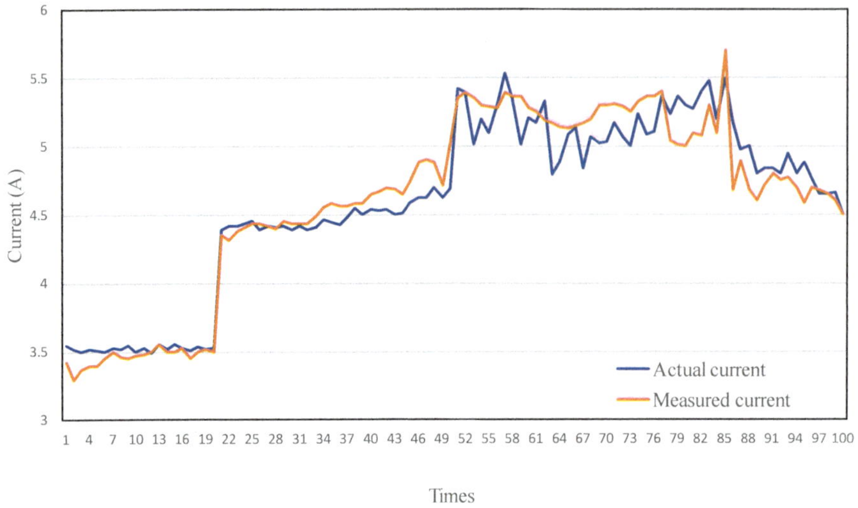

Fig. 6. Current sensor evaluation

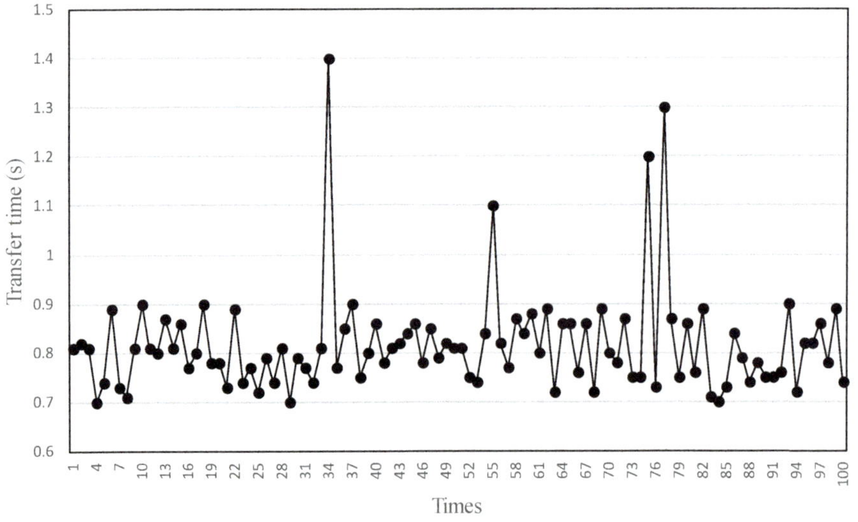

Fig. 7. Power consumption data transfer

4 Conclusion

In this work, the prototype of DC Smart Plug system could be applied to efficiently control the electrical appliances through wireless signal. The design of the data transfer format was crucial for the development of the smart plug. When the electronic board was connected directly to the database, the hardware became overworked due to simultaneous multi-function processing. The usage of Cloud Server provided more efficient operation from the reduction of the electronic board processing. The measurement and display of power consumption data assisted the smart plug calculations and appliance control.

Acknowledgement. The authors gratefully acknowledge the support provided by the NICOP Grant from the Office of Naval Research, USA for the development of the Community PV DC Micro-grid System in the Smart Community and Asian Development College for Community Economy and Technology (adiCET), Chiang Mai Rajabhat University for providing analysis tools for this research.

References

1. Ding, Z., Wang, G., Liu, Z., Long, R.: Research on differences in the factors influencing the energy-saving behavior of urban and rural residents in China–a case study of Jiangsu Province. Energy Policy **100**, 252–259 (2017)
2. Shakeri, M., Shayestegan, M., Abunima, H., Reza, S.M.S., Akhtaruzzaman, M., Alamoud, A.R.M., Sopian, K., Amin, N.: An intelligent system architecture in home energy management systems (HEMS) for efficient demand response in smart grid. Energy Build. **138**, 154–164 (2017)
3. Szcześniak, P., Kaniewski, J.: Power electronics converters without DC energy storage in the future electrical power network. Electr. Power Syst. Res. **129**, 194–207 (2015)
4. Sasidharan, N., Madhu M, N., Singh, J.G., Ongsakul, W.: An approach for an efficient hybrid AC/DC solar powered Homegrid system based on the load characteristics of home appliances. Energy Build. **108**, 23–35 (2015)
5. Tang, S., Kalavally, V., Ng, K.Y., Parkkinen, J.: Development of a prototype smart home intelligent lighting control architecture using sensors onboard a mobile computing system. Energy Build. **138**, 368–376 (2017)
6. Risteska Stojkoska, B.L., Trivodaliev, K.V.: A review of Internet of Things for smart home: Challenges and solutions. J. Clean. Prod. **140**(Part 3), 1454–1464 (2017)
7. Mohammed Shahanas, K., Bagavathi Sivakumar, P.: Framework for a Smart water management system in the context of smart city initiatives in India. Procedia Comput. Sci. **92**, 142–147 (2016)
8. Leonardi, A., Ziekow, H., Strohbach, M., Kikiras, P.: Dealing with data quality in smart home environments—lessons learned from a smart grid pilot. J. Sens. Actuat. Netw. **5**(1), 5 (2016)
9. Kazala, R., Taneva, A., Petrov, M., Penkov, S.: Wireless network for mobile robot applications. IFAC-PapersOnLine **48**(24), 231–236 (2015)
10. Silveira, E., Bonho, S.: Temperature monitoring through wireless sensor network using an 802.15.4/802.11 gateway. IFAC-PapersOnLine **49**(30), 120–125 (2016)
11. Kelly, K.E., Whitaker, J., Petty, A., Widmer, C., Dybwad, A., Sleeth, D., Martin, R., Butterfield, A.: Ambient and laboratory evaluation of a low-cost particulate matter sensor. Environ. Pollut. **221**, 491–500 (2017)
12. Chooruang, K., Mangkalakeeree, P.: Wireless heart rate monitoring system using MQTT. Procedia Comput. Sci. **86**, 160–163 (2016)
13. Abdulrahman, T.A., Isiwekpeni, O.H., Surajudeen-Bakinde, N.T., Otuoze, A.O.: Design, specification and implementation of a distributed home automation system. Procedia Comput. Sci. **94**, 473–478 (2016)

14. Piesciorovsky, E.C., Schulz, N.N.: Comparison of non-real-time and real-time simulators with relays in-the-loop for adaptive overcurrent protection. Electr. Power Syst. Res. **143**, 657–668 (2017)
15. Szabo, Z., Marcoň, P., Roubal, Z., Zezulka, F., Veselý, I., Saidl, O., Lahodny, L.: Remotely controlled smart metering for the smart home. IFAC-PapersOnLine **49**(25), 235–240 (2016)
16. Kaczmarek, M., Szatilo, T.: Reference voltage divider designed to operate with oscilloscope to enable determination of ratio error and phase displacement frequency characteristics of MV voltage transformers. Measurement **68**, 22–31 (2015)
17. Łukasz, P., Marcin, O., Dariusz, K., Marek, J.: I2C interface design for hardware master devices. IFAC Proc. Vol. **45**(7), 163–168 (2012)
18. Thaker, T.: ESP8266 based implementation of wireless sensor network with Linux based web-server. Paper presented at the 2016 Symposium on Colossal Data Analysis and Networking (CDAN) (2016)
19. Slaimi, F., Sellami, S., Boucelma, O., Hassine, A.B.: Leveraging track relationships for web service recommendation. Paper presented at the 2016 IEEE 13th International Conference on e-Business Engineering (ICEBE) (2016)

Decorrelation of Wireless Channel Coefficients for Secret Key Generation

Xiaofu Wu(✉), Dongming Dai, Xunjian Yu, and Jun Yan

Nanjing University of Posts and Telecommunications, Nanjing 210003, China
{xfuwu,yanj}@njupt.edu.cn, 563135327@qq.com, yu_xunjian@163.com

Abstract. When Alice and Bob can observe a common wireless radio channel, it is possible to extract secrets between them. As the wireless channel coefficients over time are generally correlated, it is a common challenge to perform decorrelation efficiently for various secret key generation schemes. In the literature, it was reported that the Karhunen-Loève (K-L) transform for decorrelation is very sensitive in implementation and often the channel covariance at the side of Alice should be finely transferred to Bob, which is very expensive. In this paper, we provide some insights on the sensitivity of K-L transform, which can be well exploited to develop an improved version of K-L transform for removing this sensitivity. Since the K-L transform is very expensive in complexity, we further propose a linear prediction self-filtering approach at both sides for decorrelation. The theoretical analysis show that its performance is insensitive to the impairment of the wireless channel coefficients observed at different sides, which is further validated by simulations.

Keywords: Secret key generation · Reconciliation
Karhunen-Loève transform · Linear prediction filtering

1 Introduction

Common randomness available at Alice and Bob can be exploited for secure communications between them [1,2], in which a secret key can be explicitly extracted. It is soon recognized that the reciprocal wireless channel in time-division duplexing (TDD) mode can be well employed as common randomness and secret key extraction from wireless channel coefficients is possible [3].

To extract secret key from wireless channel coefficients, it requires three phases in general. (1) Channel sensing and quantization: Alice and Bob each measure the channel impulse response (CIR). With quantization, the measurements are converted into a string of binary bits. (2) Reconciliation: The bit discrepancies at the both sides are corrected with an information reconciliation process through public discussion. (3) Privacy amplification: The reconciliated bit streams are further compressed by use of universal hashing techniques, and the resulting compressed bit string (a key) should be uniformly-distributed and completely-unpredictable from any eavesdroppers.

S. O. Hwang et al. (Eds.): ICGHIT 2018, LNEE 502, pp. 149–161, 2019.
https://doi.org/10.1007/978-981-13-0311-1_26

In the past years, various approaches have been proposed for the generation of secret key from wireless channels [4–11]. In [4,8], Mathur et al. introduced a level-crossing algorithm. It can be efficiently implemented, which, however, is often with relatively low rate for generating secret key. In [9–11], multi-bit quantization methods are proposed to handle this shortage, which, however, suffer from high probability of disagreement for generated key bits at the both sides. The information reconciliation procedure has been introduced in [9,10] to reduce the bit disagreement, and an iterative interactive Cascade protocol was employed in [11].

As an ingredient step for secret key generation, privacy amplification is to remove the redundancy of bit strings. The redundancy comes from both the correlation of channel coefficients over time, and the exchanged information for reconciliation which is publicly accessible for any eavesdroppers. Currently, it still remains a challenge for theoretically determining the amount of redundant information for the purpose of privacy amplification, especially for key generation from time-correlated channel coefficients.

It was reported in [9] that the direct use of K-L transform for decorrelation could be very sensitive in implementation. Indeed, whenever K-L transform is independently implemented at both Alice and Bob, the bit disagreement ratio is unusually high, which makes it doubtful in practice. In this paper, we address the sensitivity of K-L transform. Some insights are provided, which can be well exploited to develop an improved version of K-L transform for removing this sensitivity. Furthermore, both theoretical analysis and simulations show that the linear prediction filtering approach is also suitable for decorrelation in secret key generation from wireless channel coefficients.

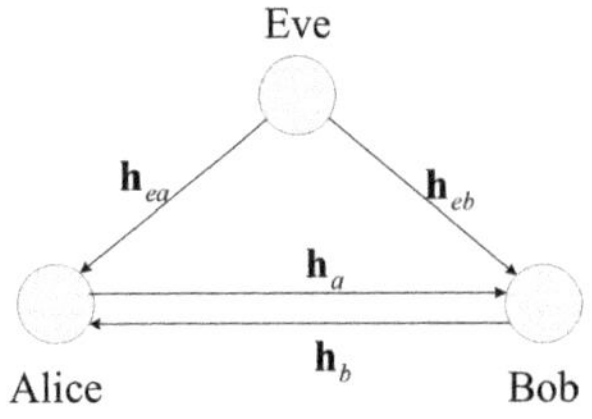

Fig. 1. Adversary model.

2 System Model

Consider the scenario in Fig. 1, where two legitimate users, Alice and Bob, want to produce a shared secret key from wireless reciprocal channels, while a passive eavesdropper, Eve, does her best to deduce information about the generated key between Alice and Bob.

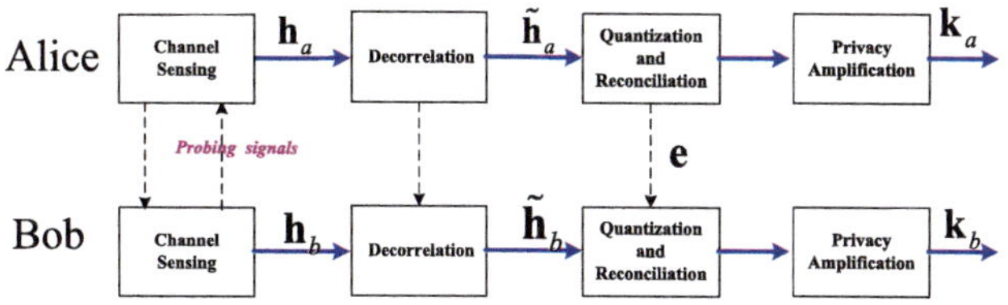

Fig. 2. The diagram of a typical secret key generation scheme

As a passive eavesdropper, Eve can estimate both channels from Alice and Bob. However, she cannot engage in an active attack against the probing process. The secret key generation method is assumed to be publicly available and the information exchanged over public discussion is also known to Eve. As Eve cannot be very close to Alice or Bob in a distance of the wavelength of the radio wave [12], it is assumed that Eve cannot obtain any information about the CIRs between Alice and Bob.

A basic secret key generation scheme with K-L transform can be shown in Fig. 2. It consists four phases. In the first phase, both Alice and Bob do channel estimation to obtain the measurements of the CIRs. Karhunen-Loève (K-L) transform is employed to decorrelate the refined channel measurements in the second phase. In the third phase, quantization and reconciliation is carried out for generating common binary key bits between Alice and Bob [4,8–11,13–15]. Finally, privacy amplification is performed to remove the key redundancy.

To evaluate the proposed method, three metrics are of primary interests.

- Secret bit rate R_k: the number of independent secret bits generated per second;
- Bit disagreement ratio P_{BD}: the probability that a single bit generated by Alice and Bob is not identical;
- Randomness of generated bit: the quantity defined to ensure that the generated secret bits are random and unpredictable, which is tested in this paper using NIST test suite [16].

3 Channel Decorrelation for Secret Key Generation

3.1 Channel Sensing

Let $\mathbf{h}$ denote the true CIR between Alice and Bob. By sending probes, both Alice and Bob can obtain imperfect channel estimates of $\mathbf{h}$, namely,

$$\mathbf{h}_a = \mathbf{h} + \mathbf{z}_a \tag{1}$$

$$\mathbf{h}_b = \mathbf{h} + \mathbf{z}_b \tag{2}$$

where $\mathbf{z}_a$ and $\mathbf{z}_b$ represent zero-mean random Gaussian-distributed variables due to imperfect estimation at the sides of Alice and Bob, respectively. In this paper, we consider Rayleigh fading channels. Hence, $\mathbf{h}$, $\mathbf{z}_a$ and $\mathbf{z}_b$ are three independent Gaussian variables, namely, $\mathcal{N}\left(0,\sigma^2\right)$, $\mathcal{N}\left(0,\sigma_a^2\right)$ and $\mathcal{N}\left(0,\sigma_b^2\right)$, respectively. Here, it is assumed that $\sigma_a^2=\sigma_b^2$. The signal-to-noise ratio (SNR) can be defined as $SNR=\frac{\sigma^2}{\sigma_a^2}=\frac{\sigma^2}{\sigma_b^2}$.

3.2 Level-Crossing-Based Filtering

In [4,8], the level-crossing algorithm is employed to generate binary bit string for extracting secret key, and also the timing indexes of excursions for information reconciliation, by using excursions in the channel measurements as inputs. A favorite feature of the level-crossing algorithm is that information reconciliation can be performed via the simple exchange of positions of excursions, which does not leakage any useful information for the generated key. Hence, privacy amplification is not required.

Here, we employ the level-crossing algorithm for locating *reliable* channel measurements at the both sides. With channel estimation, both Alice and Bob can get a sequence of n channel measurements, denoted as $\mathbf{h}_a=\{h_a(1),h_a(2)\cdots h_a(n)\}$ and $\mathbf{h}_b=\{h_b(1),h_b(2)\cdots h_b(n)\}$, respectively. Let q_+,q_- be two thresholds for locating excursions in both $\mathbf{h}_a$ and $\mathbf{h}_b$

$$q_+=mean(U^n)+\alpha\cdot\sigma(U^n) \tag{3}$$
$$q_-=mean(U^n)-\alpha\cdot\sigma(U^n) \tag{4}$$

where $U^n\in\{\mathbf{h}_a,\mathbf{h}_a\}$, $\sigma(\cdot)$ is the standard deviation and α is set to control the thresholds.

Accumulated level-crossing algorithm is stated as follows.

1. Alice parses through her channel estimates to find excursions in which m or more successive estimates are entirely above q_+ or below q_- (m is a protocol parameter to denote the minimum number of estimates in a excursion that we wish to seek).
2. For each excursion found in Step 1, Alice sends to Bob its location, which is defined as the timing index of the channel estimate lying in the center of the excursion. Locations of all such excursions are collected as a list $\mathcal{L}$.
3. Bob gets $\mathcal{L}$ from Alice by public discussion. For each index of $\mathcal{L}$, Bob checks to see if $m-1$ channel estimates centered around this index are all above q_+ or below q_-. If this check passes, the index will be included in a new list $\tilde{\mathcal{L}}$; otherwise, the index will be abandoned. Finally, Bob sends $\tilde{\mathcal{L}}$ to Alice through the public channel.
4. For each position $i\in\tilde{\mathcal{L}}$, $m-1$ channel estimates centered around position i are accumulated at the both sides as

$$s_a(i)=\sum_j h_a(j),s_b(i)=\sum_j h_b(j)$$

for $j=\tilde{\mathcal{L}}(i)-\lfloor\frac{m-2}{2}\rfloor,\cdots,\tilde{\mathcal{L}}(i)+\lceil\frac{m-2}{2}\rceil$, $i=1,2,\cdots,|\tilde{\mathcal{L}}|$.

In the end of this phase, Alice and Bob can obtain the refined signal vector $\mathbf{s}_a$ and $\mathbf{s}_b$, respectively. In contrast with $\mathbf{h}_a$ and $\mathbf{h}_b$, $\mathbf{s}_a$ and $\mathbf{s}_b$ are of higher *SNR*, but of reduced size. Clearly, this property makes them suitable for subsequent multi-bit quantization, possibly with a higher secret bit rate R_k and a lower bit disagreement ratio P_{BD}, compared to the traditional level-crossing scheme.

3.3 Decorrelation

In general, the channel coefficients sampled at different time epoches are correlated. To generate a secret key, it is desirable to de-correlate these channel coefficients for subsequent quantization and reconciliation. Hence, there are two main challenges for decorrelation. One is to ensure that the underlying variables are uncorrelated before quantization, and the other is to facilitate the process of reconciliation.

Consider that Alice and Bob have got their estimates of the wireless channel, namely, $\mathbf{h}_a$ and $\mathbf{h}_b$. Then, decorrelation should be performed at both sides of Alice and Bob. Mathematically, decorrelation at both sides can be seen as two mapping

$$\begin{aligned}\tilde{\mathbf{h}}_a &= \mathcal{L}_a(\mathbf{h}_a),\\ \tilde{\mathbf{h}}_b &= \mathcal{L}_b(\mathbf{h}_b).\end{aligned} \tag{5}$$

which maps $\mathbf{h}$ into $\tilde{\mathbf{h}}$ of the same size. For tractability, both $\mathcal{F}_a(\cdot)$ and $\mathcal{F}_b(\cdot)$ are assumed to be linear. With decorrelation, the output sequence $\tilde{\mathbf{h}} = [\tilde{h}_1, \cdots, \tilde{h}_N]$ should be uncorrelated over time, namely, $E\{\tilde{h}_i^* \tilde{h}_j\} = 0$ for $\forall i \neq j$.

Then, the output sequence $\tilde{\mathbf{h}}$ is individually quantized to obtain

$$\begin{aligned}\mathbf{k}_a &= \mathcal{Q}(\mathbf{h}_a),\\ \mathbf{k}_b &= \mathcal{Q}(\mathbf{h}_b).\end{aligned} \tag{6}$$

As $\mathbf{k}_a$ may be different with $\mathbf{k}_b$, reconciliation is an ingredient step for getting a common key at the both sides, in which, Alice and Bob exchange information to correct the discrepancy between their individual key. Clearly, the amount of information exchanged should be minimized.

By concatenating (7) with (5), we have that

$$\begin{aligned}\mathbf{k}_a &= \mathcal{Q} \circ \mathcal{L}_a(\tilde{\mathbf{h}}_a),\\ \mathbf{k}_b &= \mathcal{Q} \circ \mathcal{L}_b(\tilde{\mathbf{h}}_b).\end{aligned} \tag{7}$$

To minimize the amount of information exchanged for the purpose of reconciliation, it is reasonable to choose two linear mapping $\mathcal{L}_a$ and $\mathcal{L}_b$ for minimizing the Euclidian distance between $\mathcal{L}_a(\tilde{\mathbf{h}}_a)$ and $\mathcal{L}_b(\tilde{\mathbf{h}}_b)$, namely,

$$\begin{aligned}&\min_{\mathcal{L}_a,\mathcal{L}_b} E_{\mathbf{h},\mathbf{z}_a,\mathbf{z}_b} \left|\mathcal{L}_a(\tilde{\mathbf{h}}_a) - \mathcal{L}_b(\tilde{\mathbf{h}}_b)\right|^2\\ &= \min_{\mathcal{L}_a,\mathcal{L}_b} E_{\mathbf{h},\mathbf{z}_a,\mathbf{z}_b} \left|\mathcal{L}_a(\tilde{\mathbf{h}}) - \mathcal{L}_b(\tilde{\mathbf{h}}) + \mathcal{L}_a(\mathbf{z}_a) - \mathcal{L}_b(\mathbf{z}_b)\right|^2.\end{aligned}$$

For linear transforms $\mathcal{L}_a, \mathcal{L}_b$ with the purpose of decorrelation, we can consider equivalent unitary matrix forms of U_a, U_b of size $N \times N$, as demonstrated by the K-L transforms shown later in this section. Let $\Delta U = U_a - U_b$, we have that

$$\begin{aligned}
&\min_{\mathcal{L}_a,\mathcal{L}_b} E_{\mathbf{h},\mathbf{z}_a,\mathbf{z}_b} \left|\mathcal{L}_a(\tilde{\mathbf{h}}_a) - \mathcal{L}_b(\tilde{\mathbf{h}}_b)\right|^2 \\
&= \min_{U_a,U_b} E_{\mathbf{h},\mathbf{z}_a,\mathbf{z}_b} \left|U_a\tilde{\mathbf{h}}_a - U_b\tilde{\mathbf{h}}_b\right|^2 \\
&= \min_{U_a,U_b} E_{\mathbf{h},\mathbf{z}_a,\mathbf{z}_b} \left|\Delta U \cdot \mathbf{h} + U_b(\mathbf{z}_a - \mathbf{z}_b) + \Delta U \cdot \mathbf{z}_a\right|^2 .
\end{aligned}$$

Since

$$\begin{aligned}
&E_{\mathbf{h},\mathbf{z}_a,\mathbf{z}_b} \left|\Delta U \cdot \mathbf{h} + U_b(\mathbf{z}_a - \mathbf{z}_b) + \Delta U \cdot \mathbf{z}_a\right|^2 \\
&\leq |\Delta U|_2^2 \cdot E_{\mathbf{h}}|\mathbf{h}|_2^2 + |U_b|_2^2 \cdot E_{\mathbf{z}_a,\mathbf{z}_b}|\mathbf{z}_a - \mathbf{z}_b)|_2^2 \\
&+|\Delta U|_2^2 \cdot E_{\mathbf{z}_a}|\mathbf{z}_a|_2^2 \\
&= |\Delta U|_2^2 \cdot \sigma^2 + |U_b|_2^2 \cdot 2\sigma_a^2 + |\Delta U|_2^2 \cdot \sigma_a^2 \\
&= |\Delta U|_2^2 \cdot (\sigma^2 + \sigma_a^2) + 2\sigma_a^2,
\end{aligned}$$

a suboptimum criterion for minimizing the amount of information exchanged for key reconciliation is

$$\min_{U_a,U_b} |U_a - U_b|_2^2. \tag{8}$$

3.4 K-L Transform

Theoretically, the decorrelation can be done with the K-L transform [9]. Without loss of generality, let $\mathbf{h} = [h_1, \cdots, h_N]^T$ be the input random sequence of length N drawn from a wide-sense stationary process of mean $\mathbf{v}$ and covariance matrix R. Let $\mathbf{u}_1, \cdots, \mathbf{u}_N$ be eigenvectors associated with N eigenvalues of the matrix R. With K-L transform, $\mathbf{h}$ can be expanded as a linear combinations of N eigenvectors

$$\mathbf{h} = \sum_{i=1}^{N} y_i \mathbf{u}_i. \tag{9}$$

where the coefficients $\{y_i\}_{i=1}^N$ are uncorrelated random variables defined by the inner product

$$y_i = \mathbf{u}^\dagger \mathbf{h}. \tag{10}$$

Collectively, $\mathbf{y} = [y_1, \cdots, y_N]^T$ can be written as

$$\mathbf{y} = U^\dagger \mathbf{h}, \tag{11}$$

where $U = [\mathbf{u}_1, \cdots, \mathbf{u}_N]$ is the eigenvector matrix of R and

$$R = U^\dagger \Lambda U. \tag{12}$$

with Λ denoting the diagonal matrix of eigenvalues with $\Lambda_{ii} = \lambda_i$.

The potential problem for the K-L transform is that the eigenvalue decomposition of a Toeplitz matrix is not unique. Indeed, if $\mathbf{u}_i$ is an eigenvector of R, $-\mathbf{u}_i$ is also an eigenvector. Hence, any form of unitary matrix $U = [\pm\mathbf{u}_1, \cdots, \pm\mathbf{u}_N]$ is also an eligible eigenvector matrix of R. For the numerical computation of eigenvalue decomposition for a given matrix R, the eigenvector matrix U is not unique in general. Hence, it may result into significant discrepancy (8) between Alice and Bob, which is unacceptance for the purpose of reconciliation.

Consider that Alice and Bob respectively estimate the covariances R_a and R_b from their channel estimates $\mathbf{h}_a$ and $\mathbf{h}_b$, perform the necessary eigen-decomposition, and construct the independent channel vector

$$\hat{\mathbf{h}}_\xi = U_\xi^\dagger \mathbf{h}_\xi, \tag{13}$$

where $R_\xi = U_\xi^\dagger \Lambda_\xi U_\xi$ and $\xi \in a, b$.

When Alice and Bob each decorrelate using their own estimates of the covariance. As shown in [10], the bit-disagreement-ratio performance is very poor under these circumstances. If, on the other hand, Alice estimates the covariance and transmits the resulting eigenvectors to Bob for use in the decorrelation process, the performance can be improved dramatically.

3.5 Self-reconciliation with K-L Transform

For independent estimation of the covariance matrix, the calculated eigenvectors at both Alice and Bob are very similar whenever $|R_a - R_b|$ is small, which is the case due to the channel reciprocity. The main problem is that they may differ in the vector direction, which means that it can be frequently observed for $|\mathbf{u}_1^a - \mathbf{u}_1^b| \approx 2|\mathbf{u}_1^a|$ instead of $|\mathbf{u}_1^a - \mathbf{u}_1^b| \approx 0$. Hence, one can do simple negotiation for reconciliation of direction of eigenvectors as follows.

Let $\mathbf{u}_\xi = [u_1^\xi, u_2^\xi, \cdots, u_N^\xi]^T, \xi \in \{a, b\}$ be one of the eigenvectors computed from R_ξ. To achieve the agreement of eigenvectors at both sides, a straightforward method is to employ a refined form of $\mathbf{u}_\xi$ at both sides as

$$\mathbf{u}_\xi \to \mathrm{sign}(u_1^\xi)\mathbf{u}_\xi \triangleq \left[|u_1^\xi|, \mathrm{sign}(u_1^\xi)u_2^\xi, \cdots, \mathrm{sign}(u_1^\xi)u_N^\xi\right]^T. \tag{14}$$

In a matrix form, this can be performed as

$$U_\xi \to U_\xi \cdot \mathrm{diag}\left\{\mathrm{sign}(u_{11}^\xi), \mathrm{sign}(u_{12}^\xi), \cdots, \mathrm{sign}(u_{1N}^\xi)\right\} \tag{15}$$

at both sides of Alice and Bob independently. Hence, no public negotiation is required any more.

With this self-reconciliation approach, there is no any need for further reconciliation of eigenvector matrices between Alice and Bob, which contrasts sharply to the direct matrix transmission method [9].

4 A Linear Filtering Approach for Self Reconciliation

As the wireless channel coefficients are linearly correlated, linear filtering approach can also be well applied. In essence, the wireless channel coefficients are often modeled as autoregressive (AR) process, linear prediction can be employed to produce uncorrelated prediction errors, which can be further processed for extracting a common key between Alice and Bob.

Suppose that a linear forward predictor is of M-order. The forward prediction error can be expressed as

$$f_n = h_n + \sum_{l=1}^{M} w_l h_{n-l}, \tag{16}$$

where $w_k, k = 1, \cdots, M$ denote M tap weights. The tap weights are the solution to the augmented Wiener-Hopf equation

$$R_{M+1} \cdot \mathbf{w} = \begin{bmatrix} P_M \\ \mathbf{0} \end{bmatrix}, \tag{17}$$

where $R_{M+1} = E\left\{\mathbf{h}_{M+1}\mathbf{h}_{M+1}^{\dagger}\right\}$ with $\mathbf{h}_{M+1} = [h_0, h_1, \cdots, h_M]^T$, $\mathbf{w} = [1, w_1, \cdots, w_M]^T$ and $P_M = \sum_{l=0}^{M} w_l r_l$ denotes the prediction error power with $r_l = E\left[h_n h_{n-l}^*\right]$.

It is well-known that Levinson-Durbin algorithm can be efficiently used to solve the Wiener-Hopf equation. Essentially, it employs both forward and backward prediction with order recursion form. By including the filter order, we have to define $w_{M,l} = w_l, l = 1, \cdots, M$ and $w_{M,0} = 1$.

The tap weight of a forward prediction error filter can be order-updated as

$$w_{m,l} = w_{m,l-1} + \kappa_m w_{m-1,m-l}^*, l = 0, 1, \cdots, m \tag{18}$$

where κ_m is a constant, $w_{m,l}$ is the lth tap weight of a forward prediction-error filter of order m, and $w_{m-1,m-l}^*$ is the lth tap weight of a backward predication-error filter of order $m-1$. Note that $w_{m-1,0} = 1$ and $w_{m-1,m} = 0$. Accordingly, the tap weight of a backward prediction error filter is order-updated as

$$w_{m,m-l}^* = w_{m,l-1}^* + \kappa_m^* w_{m-1,l}. \tag{19}$$

Let $\Delta_{m-1} = \sum_{l=0}^{m-l} r_{l-m} w_{m-1,l}$ and P_m denote the prediction error power of a forward error filter of order m. Then, P_m and κ_m can be recursively computed as

$$P_m = P_{m-1} + \kappa_m \Delta_{m-1}, \tag{20}$$

$$\kappa_m = \frac{\Delta_{m-1}}{P_{m-1}}, \tag{21}$$

with $P_0 = r_0$.

With an eigenvalue decomposition of covariance matrix R_{M+1} (12), it is possible to write the solution to (17) as

$$\mathbf{w} = P_M \sum_{k=0}^{M} \lambda_k^{-1} u_{k0}^* \mathbf{u}_k \tag{22}$$

and

$$P_M = \left(\sum_{k=0}^{M} \lambda_k^{-1} \left| u_{k0} \right|^2 \right)^{-1}, \tag{23}$$

where u_{k0} is the first element of the kth eigenvector $\mathbf{u}_k$ of the correlation matrix R_{M+1} and λ_k is the kth eigenvalue of R_{M+1}.

As shown in both (22) and (23), they are insensitive to the direction of eigenvectors, which could be robust in practical independent implementation at the both sides of Alice and Bob.

5 Numerical Results

We present numerical results in this section. To generate shared secret keys, the accumulated level-crossing approach [17] is employed for quantization of channel estimate for key extraction. Essentially, it employs a 1-bit quantizer $\mathcal{Q}(\cdot)$ based on the scalars q_+ and q_- that serve as threshold levels for the quantizer, where

$$\mathcal{Q}(x) = \begin{cases} 1, & \text{if} \quad x > q_+ \\ 0, & \text{if} \quad x < q_- \\ e, & \text{otherwise} \end{cases} \tag{24}$$

with e representing an undefined state.

5.1 Accumulated Level-Crossing Algorithm

In [4,8], the level-crossing algorithm is employed to generate binary bit string for extracting secret key, and also the timing indexes of excursions for information reconciliation, by using excursions in the channel measurements as inputs. A favorite feature of the level-crossing algorithm is that information reconciliation can be performed via the simple exchange of positions of excursions, which does not leakage any useful information for the generated key. Hence, privacy amplification is not required.

Here, we employ the accumulated version of level-crossing algorithm proposed in [17] for locating *reliable* channel measurements at the both sides.

We focus on three crucial quantities of interest, namely, the secret rate R_k, the disagreement ratio P_{BD}, the randomness of generated key bits. Several controllable parameters closely related to these quantities are reference thresholds q_+, q_-, the minimum number of estimates lie in each excursion m, the frequency in which Alice and Bob probe the channel f_s, the number of bits generated per input sample l.

For the simulated fading channel, we employ Jakes' simulator [18] with the maximum Doppler frequency, f_d, to control the rate of time-variation in the channel. In simulations, the typical Doppler frequency for indoor wireless environment at the carrier frequency of 2.4 GHz is assumed to be 10 Hz.

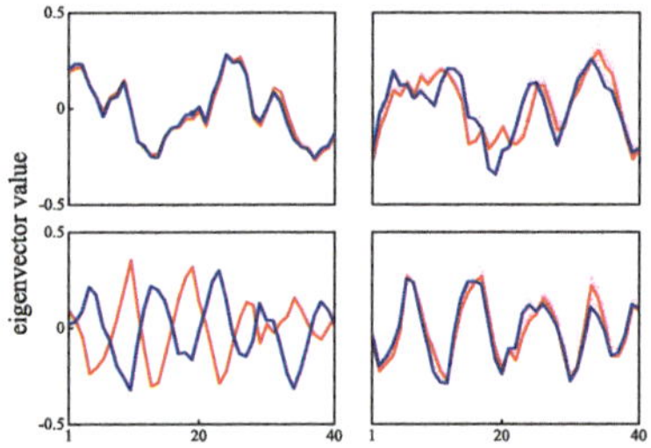

Fig. 3. First four eigenvectors of $\hat{R}_a$ and $\hat{R}_b$ with double-calculation method($m = 10, f_s = 1200, l = 2, f_d = 10\,\text{Hz}, q_{\pm} = mean \pm 0.8\sigma, N = 40, W = 20000, SNR = 18\,\text{dB}$)

5.2 Sensitivity of K-L Transform

Numerically, we find that this is due to the high sensitivity of the direction of eigenvectors for two similar covariance matrices even their difference is small. This is depicted with a tiny example shown in Fig. 3. The difference between the estimated covariance matrices, $\hat{R}_a$ and $\hat{R}_b$, can be quantitatively measured as

$$\delta R = \frac{\|\hat{R}_a - \hat{R}_b\|_F}{\|\hat{R}_a\|_F} \tag{25}$$

where $\|\cdot\|_F$ is the Frobenius norm. In this example, $\delta R = 0.2839$. However, the direction of many of eigenvectors of them could be totally different as shown in Fig. 3.

5.3 Bit Disagreement Ratio

Bit disagreement ratio is critical for the generation of the shared key between Alice and Bob. The generated bits of high P_{BD} must be discarded or corrected by exchanging large amount of reconciliation information through the public channel.

Figure 4 plots the P_{BD} as a function of the SNR for three typical reconciliation methods, namely, K-L transform without reconciliation, K-L transform with self-reconciliation and linear-filtering approach. As shown, the P_{BD} performance is very poor for K-L transform without reconciliation. The linear filtering performs very close to that of the K-L transform with self-reconciliation, but with significantly lower complexity.

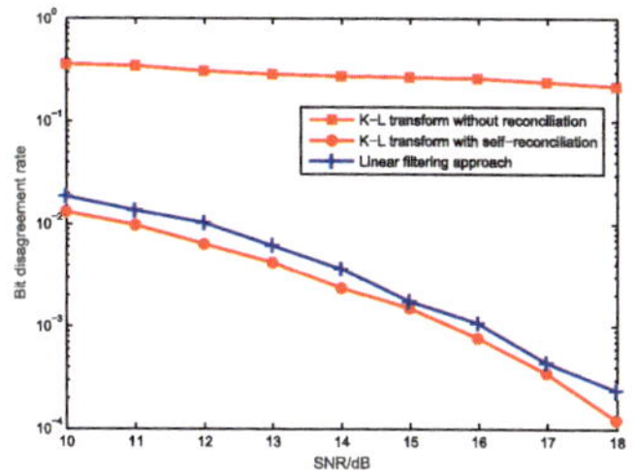

Fig. 4. Bit disagreement ratio P_{BD} for the proposed schemes ($m = 10, f_s = 1200, l = 2$, $f_d = 10\,Hz, q_{\pm} = mean \pm 0.8\sigma, N = 40, W = 20000$)

5.4 Randomness of Generated Bits

For the use of shared secret key in cryptographic applications, it must be totally random. To verify the randomness of the generated keys, we run a variety of statistical tests in the NIST Test Suite, which is a package consisting of 16 different tests [16].

These tests focus on a variety of different types of non-randomness that could exist in a key sequence. The frequency test focuses on the proportion of zeroes and ones for the entire sequence and determines whether the proportion is approximately the same as would be expected for a truly random sequence. All other tests depend on the passing of this test. Runs test focuses on the total number of runs in the sequence. Maurer's universal statistical test is devised to detect whether or not the sequence can be significantly compressed without loss of information. The approximate entropy test compares the frequency of overlapping blocks of two consecutive lengths against the expected result for a random sequence.

Each of these tests outputs a P-value which denotes that a perfect random number generator would have produced a sequence less random than the input sequence that is tested. A P-value ≥ 0.01 means that the sequence would be

Table 1. P-values of the NIST statistical tests are listed below. In each test, P-value must be more than 0.01.

Test	K-L with Self-rec	Linear filtering
Frequency	0.72	0.86
Longest run	0.42	0.39
Runs	0.64	0.71
Approx. entropy	0.82	0.77
Random excursion	0.74	0.59
Maurer's universal	0.69	0.51
Binary matrix rank	0.77	0.82
Linear complexity	0.81	0.74

considered to be random with a confidence of 99% . The results of these tests are summarized in Table 1. Note that P-values in the table are all more than 0.01, which indicates that the proposed key generation algorithm, when combined with either of two reconciliation methods does provide random secret bits without use of privacy amplification.

6 Conclusion

In this paper, we focus on the decorrelation problem for the key generation from wireless channels. Firstly, we propose a key generation algorithm, which includes channel sensing, level-crossing based accumulation, K-L transform that aims to decorrelate the channel measurements, adaptive quantization to generate bits from channel measurements. Then, we provide some insights into the sensitivity of K-L transform, which can be well exploited to develop an improved version of K-L transform for removing this sensitivity. Since the K-L transform is very expensive in complexity, we further propose a linear prediction self-filtering approach at both sides for decorrelation. Both theoretical analysis and simulations show that it performs very close to the K-L transform with self-reconciliation, but with significantly lower complexity.

Acknowledgment. This work was supported in part by the National Natural Science Foundation of China under Grant 61372123. The work of Wu was also supported by the Scientific Research Foundation of Nanjing University of Posts and Telecommunications under Grant NY213002, and by the Key University Science Research Project of Jiangsu Province under Grant 14KJA510003.

References

1. Ahlswede, R., Csiszar, I.: Common randomness in information theory and cryptography secret sharing. IEEE Trans. Info. Theory **39**, 1121–1132 (1993). https://doi.org/10.1109/18.243431
2. Maurer, U.M.: Secret key agreement by discussion from common information. IEEE Trans. Info. Theory. **39**, 733–742 (1993). https://doi.org/10.1109/18.256484
3. Hershey, E.J., Hassan, A.A., Yarlagadda, R.: Unconventional cryptographic keying variable management. IEEE Trans. Commun. **43**, 3–6 (1995). https://doi.org/10.1109/26.385951
4. Mathur, S., Trappe, W., Mandayam, N., Ye, C., Reznik, A.: Radio-telepathy: extracting a secret key from an unauthenticated wireless channel. In: Proceedings of 14th ACM International Conference Mobile Computing Network, San Francisco, California, USA (2008). https://doi.org/10.1145/1409944.1409960
5. Kitaura, A., Sumi, T., Tachibana, K., Iwai, H., Sasaoka, H.: A scheme of private key agreement based on delay profiles in UWB systems. In: Proceedings of IEEE Sarnoff Symposium, pp. 1–6, Princeton, NJ, USA (2006). https://doi.org/10.1109/SARNOF.2006.4534731
6. Wilson, R., Tse, D., Scholtz, R.: Channel identification: secret sharing using reciprocity in UWB channels. IEEE Trans. Info. Forensics Secur. **2**, 364–375 (2007). https://doi.org/10.1109/TIFS.2007.902666

7. Ye, C., Reznik, A., Shah, Y.: Extracting secrecy from jointly Gaussian random variables. In: Proceedings of IEEE International Symposium Information Theory, pp. 2593–2597, Seattle, WA, USA (2006). https://doi.org/10.1109/ISIT.2006.262101
8. Ye, C., Mathur, S., Reznik, A., Reznik, A., Shah, Y., Trappe, W., Mandayam, N.B.: Information-theoretically secret key generation for fading wireless channels. IEEE Trans. Info. Forensics Secur. **5**, 240–254 (2010). https://doi.org/10.1109/TIFS.2010.2043187
9. Patwari, N., Croft, J., Jana, S.: High-rate uncorrelated bit extraction for shared secret key generation from channel measurements. IEEE Trans. Mobile Comput. **9**, 17–30 (2009). https://doi.org/10.1109/TMC.2009.88
10. Chen, C., Jensen, A.M.: Secret key establishment using temporally and spatially correlated wireless channel coefficients. IEEE Trans. Mobile Comput. **10**, 205–215 (2011). https://doi.org/10.1109/TMC.2010.114
11. Jana, S., Premnath, N.S., Clark, M., Kasera, K.S., Patwari, N., Krishnamurthy, V.S.: On the effectiveness of secret key extraction from wireless signal strength in real environments. In: Proceedings of 15th ACM International Conference Mobile Computing Network, pp. 321–332, Beijing, China (2009). https://doi.org/10.1145/1614320.1614356
12. Durgin, G.D.: Space-Time Wireless Channels. Prentice Hall PTR, Upper Saddle River (2002)
13. Tope, M.A., McEachen, J. C.: Unconditionally secure communications over fading channels. In: Proceedings of IEEE MILCOM, pp. 54–58, McLean, VA, USA (2001). https://doi.org/10.1109/MILCOM.2001.985763
14. Chen, C., Jensen, M.A.: Improved channel quantization for secret key establishment in wireless systems. In: Proceedings of IEEE ICWITS, pp. 1–4, Honolulu, HI, USA (2010). https://doi.org/10.1109/ICWITS.2010.5611930
15. Wallace, J.W.: Secure physical layer key generation schemes: performance and information theoretic limits. In: Proceedings of IEEE International Conference on Communications, Dresden, Germany (2009). https://doi.org/10.1109/ICC.2009.5199440
16. NIST: A statistical test suite for random and pseudorandom number generators for cryptographic applications (2002). http://csrc.nist.gov
17. Cao, Z., Wu, X., Zhu, W., Cui, J.: Eliminating privacy amplification in secret key generation from wireless channels. In: Proceedings of CHINACOM 2015, pp. 266–271 (2015). https://doi.org/10.1109/CHINACOM.2015.7497949
18. Zheng, Y.R., Xiao, C.: Simulation models with correct statistical properties for Rayleigh fading channels. IEEE Trans. Commun. **51**, 920–928 (2003). https://doi.org/10.1109/TCOMM.2003.813259

Geo-Crowdsourcing and Map-Based Reporting for Smart Government

Joseph Nathanael Witanto[1] and Hyotaek Lim[2](✉)

[1] Department of Ubiquitous IT, Dongseo University, Busan 617-716, South Korea
josephwitanto@gmail.com

[2] Division of Computer Engineering, Dongseo University, Busan 617-716, South Korea
htlim@dongseo.ac.kr

Abstract. Smart government employs data collection, processing, and disseminating technology to improve services through citizen engagement and open data. Geo-crowdsourcing is a method of crowdsourcing for geospatial data, especially through smartphone application. The proposed system consists of geo-crowdsourcing mobile application for city-related events, data processing module, and map-based visualization for improving government's policy. Prototype for each module is presented and related smart city frameworks are discussed in this work.

Keywords: Crowdsourcing · Smart government · Smart city Analysis · Map visualisation

1 Introduction

A smart city employs a combination of data collection, processing, and disseminating technologies encouraging application innovation to promote the overall quality of life for its citizens and covering dimensions that include: utilities, health, transportation, entertainment and government services [2]. Technology (system and infrastructure), people (citizen engagement), and institution (government and organizations) are central components for a smart city [7]. A research of smart city roadmap development based on service, device, and technology is conducted in [6]. The roadmap served as a long-and mid-term strategic planning framework for smart city development.

Smart city government manages and implements policies through ICTs and stakeholder collaboration [9]. Smart government also utilizes data to improve their services for well-being of the community through citizen engagement and policies codevelopment [4]. One way to achieve citizen engagement in smart city is through crowdsourcing. OSM (OpenStreetMap) is example of VGI (Volunteered Geographic Information) projects which utilize crowdsourced geodata for creating open map database. OSM project also has potential of 3D VGI geo-crowdsourcing [3]. Research on [5] evaluates the usage of geo-crowdsourcing via

S. O. Hwang et al. (Eds.): ICGHIT 2018, LNEE 502, pp. 162–167, 2019.
https://doi.org/10.1007/978-981-13-0311-1_27

mobile data collection by indigenous people. People can provide location-based information using radio button, icon, camera capture, and audio recording to facilitate ecotourism asset mapping.

Geo-crowdsourcing method can be used to collect data on accident, crime, and broken public facilities so that officers can solve the problem and governments can create policies relevant to the citizens' needs. As far as the authors are aware, this is the first smart government paper with complete system from geo-crowdsourcing data collection, analysis, and map-based visualization module. The main contributions of this paper are the following.

1. Proposal of geo-crowdsourcing mobile application for city-related events.
2. Proposal of map-based reporting for various government-related activities based on real-time and past smartphone geo-crowdsourced data.
3. Evaluation of related smart city frameworks.

The remainder of the paper is organized as follows. Section 2 begins with proposed geo-crowdsourcing system. Section 3 continues with related work on smart city frameworks. Finally we conclude with future directions in Sect. 4.

2 Proposed System

Model on Fig. 1 shows the relationship between actors and proposed system. Geo-crowdsourcing mobile application receives geo-crowdsourcing data from citizen and officer (officer related with the data e.g. police officer). Data will be saved to database through CRUD (Create Read Update Delete) API in the server. After data is processed to generate analysis, the output will be visualized in map format which can be seen by government. The government can manage

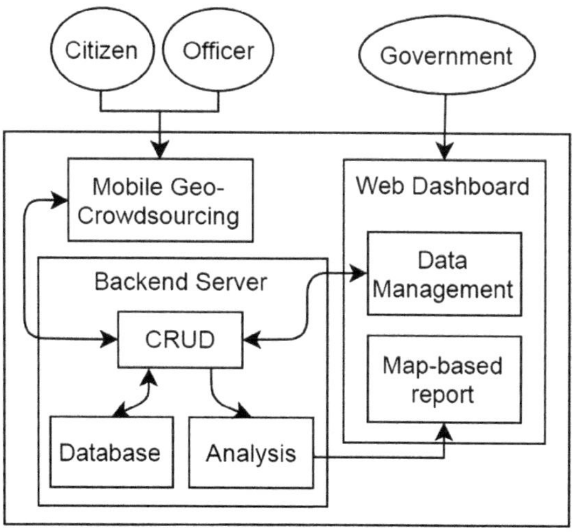

Fig. 1. Proposed system architecture

data (especially master data) in database. Geo-crowdsourcing system collects data described in Table 1. Event is the case reported by user. Type is the event category. Geoposition is the latitude and longitude of the event location. Reporting time is the time when the event happens. Area of effect is the possible radius of effect the event caused. Reporter can attach text, picture, or sound to the report.

Table 1. Geo-Crowdsourcing input

Parameter	Data
Event	Traffic accident, road potholes, etc.
Type	Accident, crime, complaint
Geoposition	GPS data
Reporting time	Timestamp
Area of effect	Area in meters
Attachment	Text, picture, sound
Trust level	1–100 (number)
Status	Ongoing, Fake, False, Resolved

Because crowdsourcing may contain fake (intentional error) or false (unintentional) error, trust level is used to rank the level of trustworthiness. Report and User (citizen and officer) have trust level. Initial value of a report's trust level depends on the reporter's trust level. After an officer checks the validity of report, the trust level may increase or decrease. Officer will have higher trust level. The citizen who used the application may register their identity card to increase their trust level. Citizen's trust level will increase or decrease depends on their past report's validity. Status represents whether the report is on the investigation, fake, false (user accidentally reported and cancelled), or resolved (report is true and case is closed).

Class diagram on Fig. 2 shows the relationship between data entities. Officer is stored as citizen who has department reference.

2.1 Data Analysis Module

Real-time data query will return report data (not aggregated) for past 24 hours. For future prediction analysis, city will be divided into $a \times b$ number of rectangular sites. Data analysis module will predict number of reports in certain site utilizing past history. Every day, resolved reports will be aggregated for every event. For every report, there will be geoposition center and radius (in meters). For all sites affected by that report, increment number of cases in that site. Every time an unresolved past data becomes resolved, past aggregated data needs to be updated. After data aggregation, for every event linear regression is done for each $a \times b$ sites to predict number of reports on the next day. This is

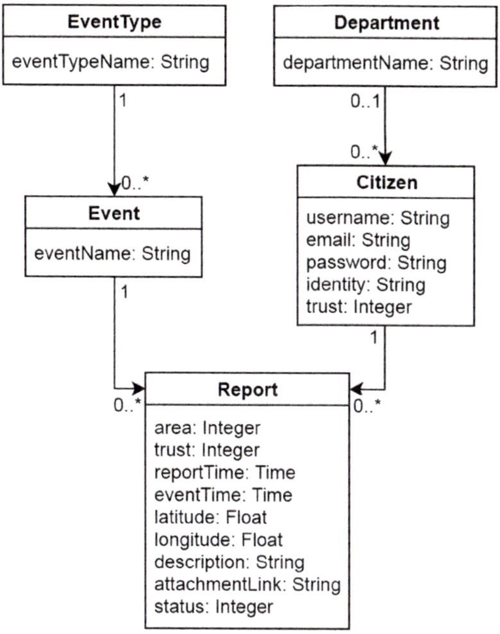

Fig. 2. Proposed class diagram

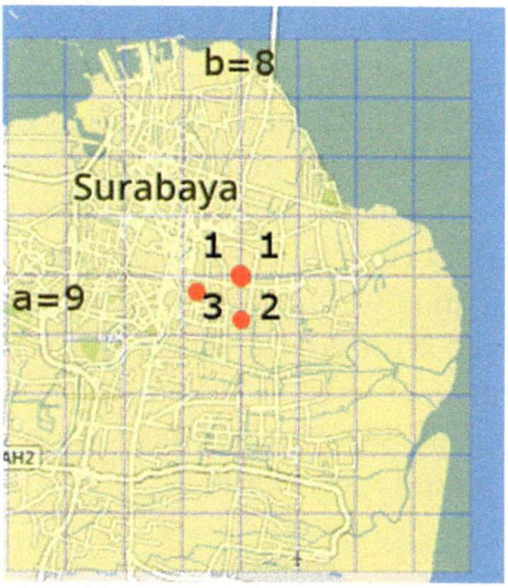

Fig. 3. Analysis aggregation example

done beforehand so when user (government) queries the process can be faster. Figure 3 shows data aggregation example for city with 9×8 sites. There are 3 reports residing in 2×2 sites. Lower left site is affected by 3 reports, lower right site is affected by 2 reports, and upper sites both are affected by 1 report.

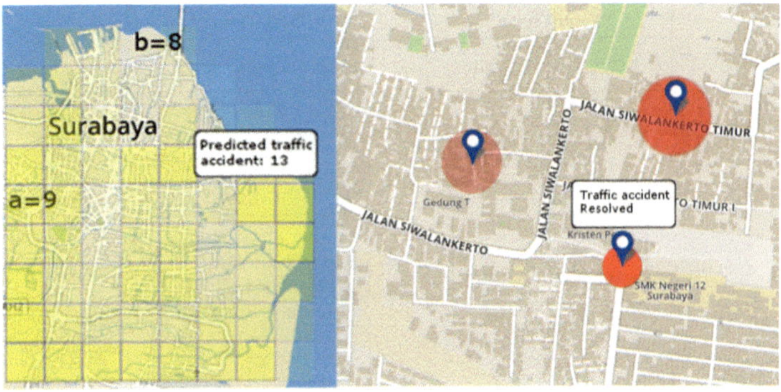

Fig. 4. Prediction map example (left) and real-time map example (right)

2.2 Map-Based Reporting Module

Map-based visualization will utilize data from OpenStreetMap overlayed with prediction or real-time data. Prediction data visualization will use output from data analysis module for next day's prediction. This can be used for government's policy or plan for citizen's safety improvement (preventive measure). Map will be overlayed with rectangles which reflect risk factor for each site. Every rectangle will have same size. Rectangle opacity will be on certain range (e.g. from 0 to 0.5), and it will reflect number of cases predicted at that site. This is achieved by normalizing number of predicted cases (dividing by maximum number of cases and multiplied by 0.5). Figure 4 (left) shows prediction map example for city with 9×8 sites. User may click the area for more detail. There are certain areas with high risk probability. Government may plan for preventive measures at certain areas to reduce the risk. Real-time data visualization will use past 24 h' report data. This can be used for monitoring (reactive measure) ongoing and resolved cases. Map will be overlayed with circles which reflect single report. Circle size depends on the report's area of effect. Circle opacity will reflect trust level of the report. Figure 4 (right) shows real-time map example at certain area. User may click the marker for more report detail.

3 Related Work

CityPulse [8] is a smart city framework that supports smart city service creation based on Internet of Things data and social media data streams (Twitter). The framework's goal is to enable cross-domain data integration. IES Cities [1] is a smart city framework that utilizes Open Government Data and sensor networks. The goal of this framework is to ease the generation of citizen-centric apps that exploit data in various domains. Different with CityPulse and IES Cities, our proposed framework utilizes crowdsourcing of geo-spatial data and presents map-based visualization specifically for city-related events.

4 Conclusion

We have discussed potential of geo-crowdsourced data for smart government. We proposed and described the system comprises of geo-crowdsourcing mobile application, backend server with analysis module, and map-based visualization report. We concluded with comparison with related smart city frameworks. We will implement and evaluate our proposed system as our future work. We may use Spring for web server development. For frontend development, we may use Angular for web dashboard, Ionic for mobile application, and Leaflet for OpenStreetMap data integration.

Acknowledgment. This work was supported by the National Research Foundation of Korea under Grant 2016R1D1A1A09916932.

References

1. Aguilera, U., Peña, O., Belmonte, O., López-de Ipiña, D.: Citizen-centric data services for smarter cities. Future Gener. Comput. Syst. **76**, 234–247 (2017)
2. Gharaibeh, A., Salahuddin, M.A., Hussini, S.J., Khreishah, A., Khalil, I., Guizani, M., Al-Fuqaha, A.: Smart cities: a survey on data management, security and enabling technologies. IEEE Commun. Surv. Tutorials **PP**(99), 1–1 (2017)
3. Goetz, M., Zipf, A.: The evolution of geo-crowdsourcing: bringing volunteered geographic information to the third dimension. In: Crowdsourcing Geographic Knowledge, pp. 139–159. Springer Netherlands, Dordrecht (2013)
4. Harsh, A., Ichalkaranje, N.: Transforming e-government to smart government: a South Australian perspective. In: Advances in Intelligent Systems and Computing, vol. 308. AISC (Volume 1), pp. 9–17. Springer, New Delhi (2015)
5. Idris, N.H., Osman, M.J., Kanniah, K.D., Idris, N.H., Ishak, M.H.I.: Engaging indigenous people as geo-crowdsourcing sensors for ecotourism mapping via mobile data collection: a case study of the Royal Belum State Park. Cartography Geogr. Inf. Sci. **44**(2), 113–127 (2017)
6. Lee, J.H., Phaal, R., Lee, S.H.: An integrated service-device-technology roadmap for smart city development. Technol. Forecast. Soc. Change **80**(2), 286–306 (2013)
7. Nam, T., Pardo, T.A.: Conceptualizing smart city with dimensions of technology, people, and institutions. In: Proceedings of the 12th Annual International Digital Government Research Conference on Digital Government Innovation in Challenging Times - dg.o 2011, p. 282 (2011)
8. Puiu, D., Barnaghi, P., Tonjes, R., Kumper, D., Ali, M.I., Mileo, A., Xavier Parreira, J., Fischer, M., Kolozali, S., Farajidavar, N., Gao, F., Iggena, T., Pham, T.L., Nechifor, C.S., Puschmann, D., Fernandes, J.: CityPulse: large scale data analytics framework for smart cities. IEEE Access **4**, 1086–1108 (2016)
9. Scholl, H.J., Scholl, M.C.: Smart governance: a roadmap for research and practice. In: iConference 2014 Proceedings, vol. 1 (2014)

The LXC-LXD Virtualization in ARM64bit X-Gene2 Server

Jin-Suk Ma(✉), Dong-Jae Kang, and Hak-Young Kim

Server Platform Res Lab, ETRI, 161 Gajeong-dong,
Yuseong-gu, Daejeon, Korea
{majinsuk,djkang,h0kim}@etri.re.kr

Abstract. The server system with low power and high efficiency ARM64bit cores is now accelerating the versatility of usages such as the low power utility, cloud computing and etc. But before the real advent using ARM64 SoC in server market, ARM64bit based server system has to satisfy the stringent operation prerequisite of the server operator. Various server virtualization techniques such as XEN, LXC-LXD, KVM-QEMU are applicable in ARM64 server with the help of world-wide developers as intel x86 based case. LXC-LXD is fully promoted by Cannonical Co. which can house the multiple applications in a single container instance differentiating the docker container. In this paper we dealt with LXC-LXD 2.0 virtualization in a real ARM64bit server reference board APM X-Gene2. The detailed process and linux kernel configuration are given in main part of this paper. Finally LXC-LXD is compared with XEN hypervisor in view of sysbench and STREAM benchmark.

Keywords: LXC-LXD · APM X-Gene2 · Server virtualization
ARM64bit

1 Introduction

The microserver system with ARM64bit server processors has been proposed recently. The microserver is a collection of server with low power and high density server processors which are composed of low power intel SoC or ARM64bit processors. One microserver rack has about 512 computing core resources and various management software within its inside with low power consumption characteristic. Hence it is now expected that microserver will make new server market area over conventional market volume which is almost governed by intel x86-64 and compatible system. Currently the performance requirement of the microserver product is satisfied by intel ATOM, Xeon-D or ARM64 server SoC. Especially in view of ARM architecture, APM(Applied Micro) is one of primary ARM server SoC and Cavium is the other one. APM has provided X-Gene1, 2 and now X-Gene3 and Cavium also done ThunderX and ThenderX2. Recently Qualcomn announced the development of Centriq 2400 SoC based on ARM core with the request of Microsoft. It is slightly known that Centriq 2400 processor exclude the conventional 32bit processing and addressing part in ARM64bit server architecture. With that exclusion the full and optimized support of 64bit architecture was achieved in Centriq 2400. With the ground of this rebased design

S. O. Hwang et al. (Eds.): ICGHIT 2018, LNEE 502, pp. 168–175, 2019.
https://doi.org/10.1007/978-981-13-0311-1_28

it results in the lightweight core, the shared L2 cache, the large amount of ring bus type L3 cache. And it also supports DDR4 6 channel interface as main memory. Currently it is estimated as counterpart of 7th generation intel processor. Apart from it, HiSilicon announced D02 server platform based on Phosphor V660 processor. According to recent report, PhosphorV660 is equipped with 32 ARM Cortex A57 cores and large size L3 cache. But its procurement is not easy outside China market. Meanwhile there are many mobile ARM processor based SoC makers such as Amlogic, Samsung, Rockchip and Broadcom. But these are out of concern in this paper because these stuffs are for mobile or embedded product not a real enterprise server. In a real data center server virtualization and management technology are not fresh trend in x86 server any more [1]. We already dealt with LXC-LXD virtualization performance with XEN with APM X-Gene1 ARM server system in our previous work [2]. XEN typical type 1 bare metal hypervisor provided an almost complete virtualization method in ARM64bit server system [7]. But it is very complicate so hard to realize for new researchers and industry practitioner. In view of linux container variants, docker container or LXC-LXD is very simple and provide a instant usage. As a result of this many real data center or cloud operators such as google and Microsoft use the docker container virtualization and the relevant orchestration scheme such as kubernetes. KVM-QEMU typical type 2 hypervisor is widely used in x86-64 server virtualization which is well-known and widely adapted. But the real execution performance of KVM-QEMU is very slow in ARM64 system because which is based on the software emulation technique inherently and on the lack of complete libvirt [3, 4]. We dealt with LXC-LXD performance measure with APM X-Gene1 in our previous work. In this paper we will deal with the newer LXC-LXD 2.0 virtualization performance with APM X-Gene2 server. LXC-LXD is a lightvisor not hypervisor as mentioned as Cannonical Co. explanation. It is similar to docker-container type virtualization at the first glance. Main difference of docker-container with LXC-LXD is that LXC-LXD allows to house many application within one internal container instance. Even if LXC-LXD has a simple virtualization scheme over the conventional XEN, the performance and virtualization functionality of it were very competitive over XEN in ARM64bit server system.

2 Development Environment

2.1 Experimental Environment

We used the APM883408-X2(code name: X-Gene2) server board as experimental hardware which was manufactured by APM. APM883408-X2 has a brand new X-Gene2 CPU based on ARM Cortex A57 8-core over a previous APM883208 X-Gene1. Main PCB layout is shown in Fig. 1 and also the hardware specification is given in Table 1. We only changed the hard drive with intel SSD 730 for getting an optimal test performance.

Fig. 1. APM883408-X-Gene2 board.

Table 1. Hardware specification

Components	Specification
Mainboard	Merlin - Applied Micro APM883408-X2-MC1
CPU	APM ARM64bit Cortex-A57 2.4 GHz 8Core
MEM	DRAM: ECC 32 GB @ 1866 MHz
HDD	Intel SSD 730 series 240 GB SATA3 (for experiment)
Boot ROM	W25Q128 Serial Flash (16 MB)
Serial port	4 port (1 console port)
USB 3.0	2 port
Gigabit Ethernet	2 port
10G Ethernet	1 port

2.2 Software Environment

Software development environment was given in Table 2, which is mainly supplied by APM with APM X-Gene2 (Code name Merlin) reference platform. In our previous work, LXC-LXD version was ver 0.27, but now is ver 2.0.2.

Table 2. Software Specification

Component	Version
Boot	xgene_sw-2.04.08-beta (Uboot-2013.04)
Kernel	xgene_sw-2.04.08-beta (Kernel 4.1.0)
Root File System	AARCH64 ubuntu 14.04.4 LTS
Cross Compiler (GCC)	gcc-linaro-aarch64-linux-gnu-4.9 or apm-aarch64-8.0.10-le (little endian)
LXC	2.0.2
LXD (commit num)	2.0.2 (9d4818a437ce8bef56b265c7b80c9530db82734d)

The disk partition information is same as our previous work and boot relevant required files are same. Please refer to above information in our previous work [2]. But the content of boot relevant files is upgraded as evolving a hardware and system kernel components. Software Specification used in this paper was shown in Table 2.

3 Setting-up Linux Kernel in LXC-LXD

Documentation and reference about LXC-LXD 2.0 can be found in [8]. LXC-LXD 2.0 is more powerful than its previous version 0.27. As it is functional, as the many kernel function is required. We encountered the many kernel configure missing messages when lxc-checkconfig command was executed. Besides of the kernel configuration about namespace support, control group support and unix98 pty support, many network function configuration in linux kernel was mandatory in LXC-LXD 2.0. We present a major kernel configuration parameter for new version in Table 3. Even if we presented the major kernel parameter setting in below Table 3, but we highly recommend our previous work to read for more understanding the process of kernel configuration in LXC-LXD.

Table 3. Kernel Setup (Major Difference Part Only)

Kernel parameters
CONFIG_NETFILTER_ADVANCED = y
CONFIG_NF_CONNTRACK = y
CONFIG_NF_CONNTRACK_IPV4 = y
CONFIG_NF_CONNTRACK_IPV6 = y
CONFIG_NF_NAT_IPV4 = y
CONFIG_NF_NAT_IPV6 = y
CONFIG_IP_NF_NAT = y
CONFIG_IP6_NF_NAT = y
CONFIG_IP_NF_TARGET_MASQUERADE = y
CONFIG_IP6_NF_TARGET_MASQUERADE = y
CONFIG_NETFILTER_XT_TARGET_CHECKSUM = y
CONFIG_CHECKPOINT_RESTORE = y
CONFIG_UNIX_DIAG = y
CONFIG_PACKET_DIAG = y
CONFIG_LEGACY_PTYS = y
CONFIG_NETLINK_DIAG = y
CONFIG_DEVKMEM = y

To install the most recent version of LXD, we cloned the LXD source code in the github, git commit number was 9d4818a437ce8bef56b265c7b80c9530db82734d. We found a README.md file in cloned root directory. LXD developers well commented about the use of this package accurately, so we could effectively realize LXC-LXD in ARM64bit server, which was mainly based on the Go programming language.

After successful installation of LXC-LXD, one can see the various command in / use/bin/ with lxc*. We installed the five ubuntu guest image (from ubunt01 to ubuntu05) to verify the functional test. LXC-LXD 2.0 has through the enormous changes in internal operation. Most of all container control environment was different in the conventional work. By referencing Stephane Graber's blog site [9], we could establish each guest image by trying below console command.

```
#sudo lxc-checkcofig
#lxc launch ubuntu:14.04 ubuntu01
#lxc list
#lxc config set ubuntu01 limits.cpu 4
#lxc config set ubuntu01 limits.memory 4096MB
#lxc config show ubuntu01
#lxc exec ubuntu01 /bin/bash
```

And also, in LXC-LXD 2.0, there is a change in web panel interface. So whoever want to use this interface have to update the remote connection part because this was not functional properly on the basis of using the previous ver 0.27 interface [9]. We also adapted this new web interface but its outlook was not attractive and fully functional at the time of writing, so we can't help recommending the console interface to interact with LXC-LXD container.

4 Experimental Results

To keep the consistency of our previous work, we sustained all testing environments with same benchmark conditions. LXC-LXD virtualization is based on a linux kernel technology such as a namespace, chroot, cgroups(control groups) and etc. So it is basically different with XEN hypervisor based architecture grasping whole hardware resources. But in view of the user sight, two of those after connecting to guest domain by the user console have nearly same behavior although XEN guest actually go through the additional linux kernel boot process. So the real performance measure in user space is comparable and affordable albeit the noncoincidence in system architecture facet exists. Below benchmark results are all updated in APM X-Gene2 platform at the time of writing. The CPU benchmark results were shown in from Tables 4, 5, 6 and 7. From Tables 4 and 5 are in case of LXC-LXD. In all tables, LXC- means LXC-LXD to simplify the table content. From Tables 6 and 7 are in case of XEN. From Tables 4, 5, 6 and 7 are the results of the condition on cpu-max-prime with 2000 and 20000 each. Finally we added STREAM mirco benchmark results in Fig. 2. In LXC-LXD, LXC-HOST has 8 CPU core, 32 GB main memory and LXC-GUEST has 4 core CPU, 4 GB memory as explained in previous chapter. As a result of these benchmark test, LXC-HOST and LXC-GUEST show almost same performance. In case of XEN, we could verify almost same CPU performance between XEN-DOM0 and XEN-DOMu. In both cases, the total time is reduced to half in 2 threads over 1 thread as expected. The disk

IO benchmark results were shown in Table 8. From Table 8, one can verify that LXC-LXD is the better disk IO performance over XEN. Especially XEN-DOMu showed a degraded performance over XEN-DOM0, which was mainly due to the virtualization overhead in XEN. The implementation of LXC-LXD in ARM64bit requires the less efforts over XEN. But it showed almost same performance and flexibility of system resource usability over XEN. With the help of STREAM macro benchmark as shown in Fig. 2, we could verify that the performances of LXC-LXD about all performance factors in STREAM such as copy, scale, add and triad were superior than XEN case.

Table 4. Maximum prime number check in CPU test (LXC-LXD, 1 threads)

		LXC-HOST	LXC-GUEST (4core, 4G)	LXC-HOST	LXC-GUEST (4core, 4G)
CPU-max-prime		**2000**		**20000**	
Total time		1.2188 s	1.2177 s	25.7519 s	25.7410 s
Total number of events		10000	10000	10000	10000
Total time taken by event execution		1.2169	1.2156	25.7499	25.7389
Per-request statistics	Min	0.12 ms	2.57 m	2.57 ms	0.12 ms
	Avg	0.12 ms	2.57 m	2.57 ms	0.12 ms
	Max	0.16 ms	2.72 ms	2.65 ms	0.16 ms
	Approx. 95 percentile	0.12 ms	2.58 ms	2.58 ms	0.12 ms

Table 5. Maximum prime number check in CPU test (LXC-LXD, 2 threads)

		LXC-HOST	LXC-GUEST (4core, 4G)	LXC-HOST	LXC-GUEST (4core, 4G)
CPU-max-prime		**2000**		**20000**	
Total time		0.6146 s	0.61 s	12.8850 s	12.8731 s
Total number of events		10000	10000	10000	10000
Total time taken by event execution		1.2181	1.2168	25.7584	25.7419
Per-request statistics	Min	0.12 ms	2.57 m	2.57 ms	0.12 ms
	Avg	0.12 ms	2.58 m	2.57 ms	0.12 ms
	Max	0.18 ms	2.81 ms	4.31 ms	0.22 ms
	Approx. 95 percentile	0.12 ms	2.59 ms	2.58 ms	0.12 ms

Table 6. Maximum prime number check in CPU test (XEN, 1 threads)

		XEN-DOM0 (2core, 2G)	XEN-DOMu (4core, 4G)	XEN-DOM0 (2core, 2G)	XEN-DOMu (4core, 4G)
CPU-max-prime		**2000**		**20000**	
Total time		1.2196 s	1.2193 s	25.7643 s	25.7567 s
Total number of events		10000	10000	10000	10000
Total time taken by event execution		1.2177	1.2174	25.7620	25.7547
Per-request statistics	Min	0.12 ms	2.57 m	2.57 ms	0.12 ms
	Avg	0.12 ms	2.58 m	2.58 ms	0.12 ms
	Max	0.21 ms	3.05 ms	2.82 ms	0.16 ms
	Approx. 95 percentile	0.12 ms	2.58 ms	2.58 ms	0.12 ms

Table 7. Maximum prime number check in CPU test (XEN, 2 threads)

		XEN-DOM0 (2core, 2G)	XEN-DOMu (4core, 4G)	XEN-DOM0 (2core, 2G)	XEN-DOMu (4core, 4G)
CPU-max-prime		**2000**		**20000**	
Total time		0.6157 s	0.6105 s	12.9247 s	12.9068 s
Total number of events		10000	10000	10000	10000
Total time taken by event execution		1.2193	1.2179	25.8466	25.8054
Per-request statistics	Min	0.12 ms	2.57 m	2.57 ms	0.12 ms
	Avg	0.12 ms	2.58 m	2.58 ms	0.13 ms
	Max	0.18 ms	4.98 ms	7.09 ms	0.37 ms
	Approx. 95 percentile	0.12 ms	2.61 ms	2.59 ms	0.12 ms

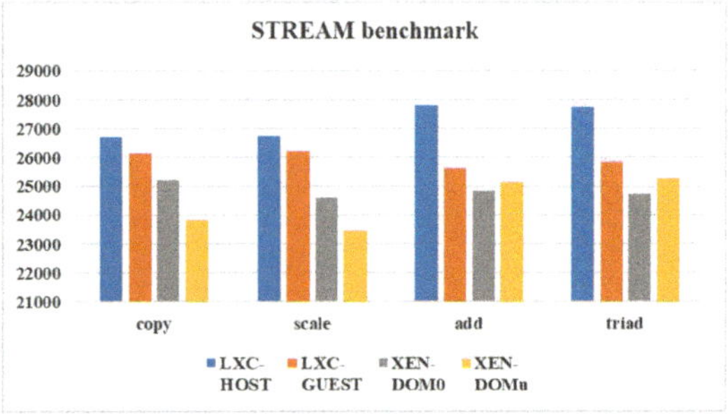

Fig. 2. STREAM benchmark results.

Table 8. Disk IO test (run-size:1G, LXC-LXD)

		LXC-HOST	LXC-GUEST (4core, 4G)	XEN-DOM0 (2core, 2G)	XEN-DOMu (4core, 4G)
Transfer rate(Mb/sec)		100.67	101.2	90.676	70.797
Total time		300.0043 s	300.0001 s	300.0038 s	300.0004 s
Total number of events		1932800	1943000	1741000	1359300
Total time taken by event execution		17.0367	17.0324	17.1080	14.3234
Per-Request Statistics	Min	0.00 m	0.00 m	0.00 ms	0.00 ms
	Avg	0.01 m	0.01 m	0.01 ms	0.01 ms
	Max	0.49 ms	6.95 ms	5.70 ms	0.25 ms
	Approx. 95 percentile	0.01 ms	0.01 ms	0.01 ms	0.01 ms

5 Conclusion

In this paper, we commented the current ARM64bit server market trend and virtualization technology. After that we implemented LXC-LXD 2.0 virtualization in a real ARM64bit server based on APM X-Gene2. In view of the server virtualization, although the other hypervisor based virtualization such as XEN or KVM-QEMU has been ready for long time, LXC-LXD or other container based technology is now predominant due to structural simplicity, operational performance, easy of use and etc. As the result of performance benchmark with LXC-LXD 2.0 and XEN, we verified that LXC-LXD was a good candidate for ARM64bit server virtualization.

References

1. Gupta, S., Gera, D.: A comparison of LXD, Docker and Virtual Machine. Int. J. Sci. Eng. Res. **7**, September 2016
2. Ma, J.S., Kim, Y.W., Choi, W.: The virtualization and performance comparison with LXC-LXD in ARM64bit Server. In: Proceedings of ICITCS 2016, Prague, Czech (2016)
3. Ma, J.S., Kim, H.Y., Choi, W.: KVM-QEMU virtualization with ARM64bit server system. In: Proceedings of CloudComp 2015 EAI Conference, Deajeon, Korea, pp. 339–348 (2015)
4. Ma, J.S., Kim, H.Y., Choi, W.: The performance enhancement of KVM-QEMU virtualization with ARM64bit server system. In: Proceedings of Global IT 2016, Hawaii, USA (2016)
5. Rosen, R.: Linux Kernel Networking: Implementation and Theory. Apress, New York
6. Arpaci-Dusseau, R.H., Arpaci-Dusseau, A.C.: Operating Systems: Three Easy Pieces. Arpaci-Dusseau Books (2015)
7. http://wiki.XEN.org/wiki/XEN_ARM_with_Virtualization_Extensions/APMXGeneMustang
8. https://insights.ubuntu.com/2016/03/22/lxd-2-0-your-first-lxd-container/
9. http://claudyus.github.io/LXC-Web-Panel/

Mobile Forged App Identification System with Centralized Signature Self-verification Method

Hyung-Woo Lee(✉) and Jaekyu Lee

Division of Computer Engineering, Hanshin University, 137, Yangsan-dong, Osan, Gyeong-gi, Republic of Korea
hwlee@hs.ac.kr, jk5309@naver.com

Abstract. Android apps developed in Java language is vulnerable to repackaging attacks as it is easy to decompile an app. Therefore, obfuscation techniques can be used to make it difficult to analyzing the source of Android apps. However, repackaging attacks are fundamentally impossible to block. Especially, it has been confirmed that most Android-based smart phones do not support verification process for the forged applications. Accordingly, the user installs and uses a fake app that appears to be functioning normally. In this case, the user is easily exposed to attacks such as leakage of personal information. Therefore, in this paper, we have constructed mobile apps identification system that applies the signature self-verification server monitoring method for Android apps and proposed a method of judging Android mobile forgery apps by performing the verification process.

Keywords: Android · Forged mobile apps · Self-verification

1 Introduction

Smartphones equipped with Android platform are spreading rapidly. Mobile app development is also growing rapidly. However, according to Android's characteristics of sharing and openness, we are performing a relatively simplified developer authentication/code checking process compared to Apple. There is a problem that the forged mobile app is easily distributed and installed through open market such that the malicious code with high security risk is hidden inside the Android application. *Googlian* malicious apps can steal Android users' Google accounts after disguising them as normal Android apps and download specific apps specified by the attacker. It also includes root privileges for Android devices and the ability to steal Google account information, causing more than a million Google accounts to leak out by the end of 2016.

Gooligan uses a more advanced attack mechanism than traditional routing techniques, so it can gain malicious root privileges on Android devices, steal Google authentication tokens and gain access to Gmail, Google Play and other key services. In addition, once the *Gooligan* malicious app is installed, malicious code hidden inside the app is activated based on modified *FakeInst, Boxer, VDLoader*, etc. [1, 2], and

S. O. Hwang et al. (Eds.): ICGHIT 2018, LNEE 502, pp. 176–182, 2019.
https://doi.org/10.1007/978-981-13-0311-1_29

Trojan horses can be attacked and the C&C server The personal information and the financial information in the mobile terminal of the user are leaked to the outside [3–5].

Therefore, this study analyzes the internal structure and operation mechanism of hidden malicious codes and provides a function to detect fake mobile apps equipped with multi-variant malicious codes linked with C&C servers. We analyzed the signature generation mechanism of the Android app, analyzed the verification process of the application installation, and developed a module that can be applied in real time for the detection of fake apps. Real-time response to the dynamic changes of the installed applications is possible and security is enhanced by self-verification. It is expected to provide reliable information protection service and secure mobile usage environment in Android environment.

2 Existing Forgery Apps Prevention Technique and Its Limitations

A program made in Java is easy to apply the decompile process. This is because it contains a lot of information in the bytecode compiled to run in the JVM. Unless it is a special case, it is mostly mapped to 1:1 and can be restored to Java code close to the original (Fig. 1).

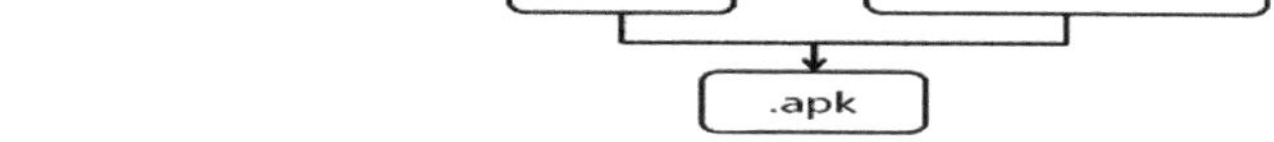

Fig. 1. APK file structure

In order to avoid source code theft or repackaging, Android Studio [6] basically provides an obfuscation tool. Changing the variable name or class name to a meaningless name changes the program logic, making it difficult for the attacker to analyze and interpret the source code. However, there is a problem that repackaging can't be completely prevented. Instead of getting the complete original source code, it does not prevent you from creating a forgery app by adding sources to your existing operating logic and repackaging it again. In the end, like the normal app, the basic behavior can provide similar malicious behavior while providing similar behavior.

Test results showed that most apps were easy to create malicious apps, and only a few apps included the ability to verify and block forged apps themselves. But this is not entirely impossible to create a forgery app. The reason is the dependency problem first. Some apps can be judged to be safe because they perform the verification process on their own, but if they include dependencies on other apps, if the app is vulnerable to forgery attacks, the problem arises that a counterfeit app can be created. The second app verification is only done and done. Do not validate newly installed applications. Most malicious applications run malware as soon as they run. Even if the execution of the malicious application is terminated, it is likely that the malicious code has already been installed.

3 Android Signature Creation and Verification Procedure

Android is similar to signing a jar file. In the META-INF folder, there are three files: MANIFEST.MF, CERT.SF, and CERT.RSA. Because APK is a kind of compressed file, it generates a signature by extracting the hash value of all the files inside the app. Therefore, it is possible to check whether the internal file of the app is forged or not through the generated hash value. The generated hash value is digitally signed with the private key and distributed with the public key, so that anyone can check the integrity of the app, thereby preventing it from being updated with the forgery app. Verification is done during installation, and the Android OS generates the hash value again and compares it with the existing signed file. Therefore, the Android platform provides a function to prevent normal apps from being updated with malicious apps. However, it does not provide a function for determining fake apps through the creation and verification of Android signatures (Fig. 2).

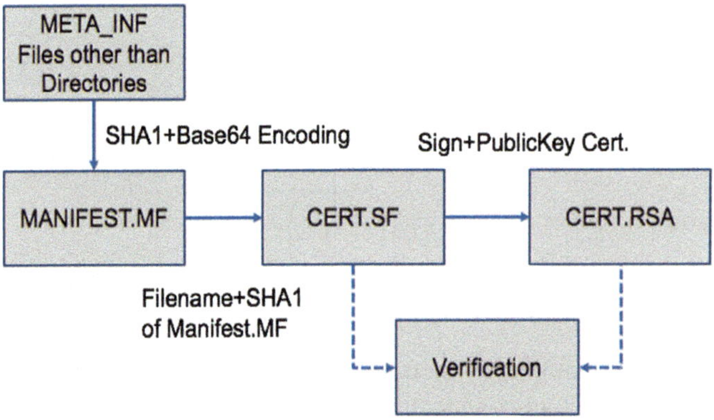

Fig. 2. Andriod signature generation & verification process

The APK file, which is an Android app, will not be installed on the terminal if a signature has not been generated or if the signature verification process fails. After generating SHA-1 based hash values for all files in the APK file, the base64 encoding process is performed. The generated hash value is created with the developer's private key and distributed with the public key certificate as shown in the Fig. 3 below. You can also verify your signature on your app through a verification process. CERT.RSA contains a public key and a value encrypted with CERT.SF as a private key. The signature value signed with the private key with the public key is decrypted and compared with the CERT.SF file to perform verification. Only the APP developer can generate legitimate signature values in APP. This function provides a function to prevent the abnormal update of the already installed APP.

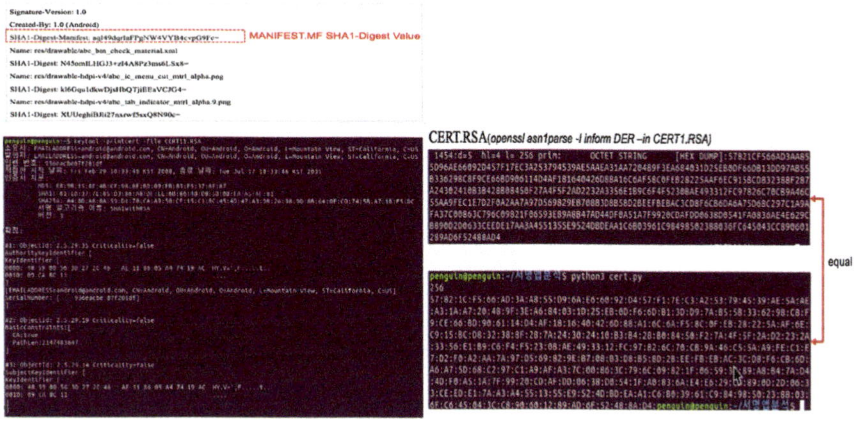

Fig. 3. Andriod signature verification example

4 Design and Implementation of Mobile Forged App Identification System

4.1 Central Monitoring Method

Applications often rely on an antivirus program, although security code is included for security reasons. This makes it easy to control and manage by performing an application that performs verification. An application sends the signature value of all applications to the verification server to check whether it is forged or not. A factor that interferes with the verification of the verification app is the disconnection of the network. In order to prevent the internal logic from changing to the FAKE server instead of the verification server, the verification server remembers whether or not each verification application is installed and the last verification time. If the last validation time passes after a certain period of validity, the user is informed of the danger through e-mail, text, or PUSH notification.

4.2 Signature Self-verification Method

When the device is booted or installed, it performs verification by not only when it is executed but also when it is installed. It also performs periodic verification. The disadvantage of the centralized control scheme is that there is a risk of losing the integrity verification of all apps if the app performing the verification is falsified. This means that if the verification app is secure, then all apps are safe. Therefore, for the integrity of the verification app, the verification is performed inside the Android OS, and the value is transmitted to the verification server again. As follow Fig. 4, we implemented mobile forgery identification(MFI) app and verification server.

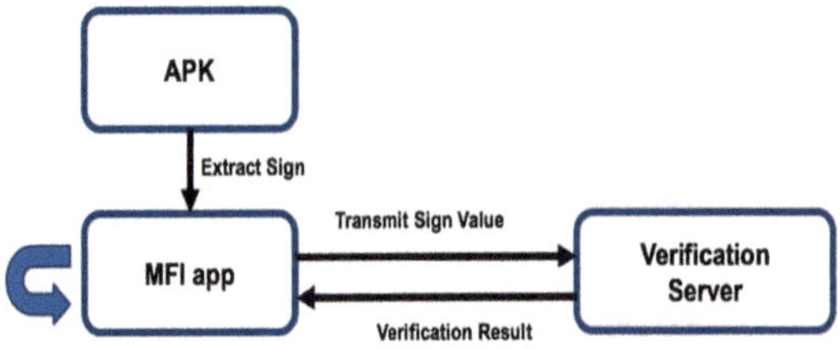

Fig. 4. Server based self-verification mechanism

4.3 Implementation and Experiments

We used android studio to create an MFI app that will be installed on a monitoring terminal to detect mobile forged apps. In addition, a validation server that provides forged app detection was built using node.js and MySQL. When the verification app starts, *selfSignService* runs in the background. The service internally verifies its own signature value at regular intervals and sends the value to the verification server. You can also invoke the list of installed packages to perform individual verification or verify all at once. The verification result will return one of three things: trust, risk, or warning. When the signature value of the application exists in the server and the result of the comparison is the same, the danger is similarly when the signature value of the application exists but the comparison result is inconsistent, the warning indicates that the normal signature value of the application is not present in the server (Figs. 5 and 6).

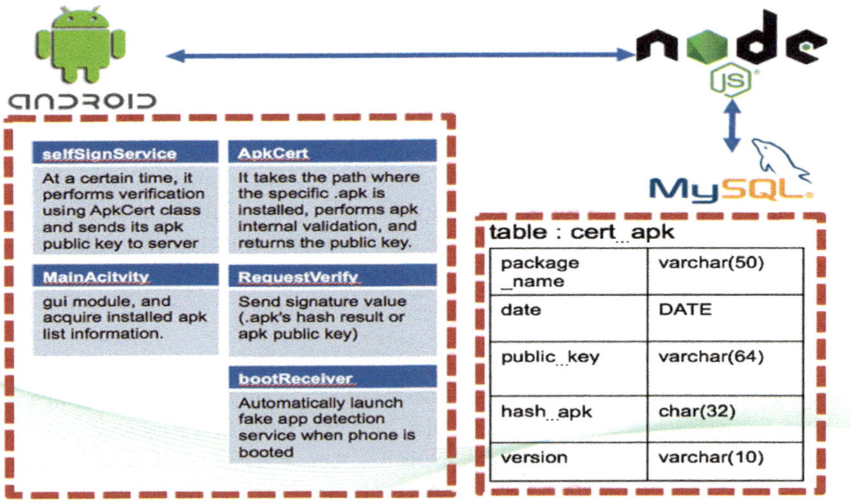

Fig. 5. Implementation of mobile forgery identification system

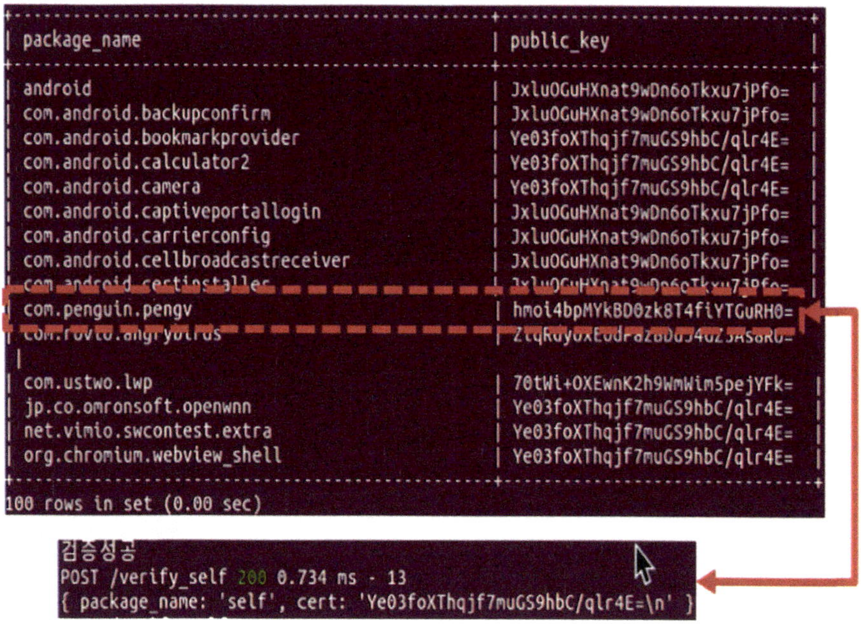

Fig. 6. Experiment result for 'pengv' apk

5 Conclusions

Since the Android application is very vulnerable to forged app production attacks, the signer has proposed a verification-based detection mechanism to solve this. The actual malicious forgery app was directly produced and tested. For most apps, it was possible to create counterfeit apps. However, some apps include logic to prevent forgery on their own. But analyzing actual malicious apps does not change the flow of normal app execution. In order to make it look normal, malicious code was added to the existing execution process, and malicious behavior was driven. Therefore, the server-based signature value self-verification method for the analysis target can provide a reliable verification process for the forgery-proofing application and can reduce the possibility of misuse detection, so that it can be applied to all mobile applications.

Acknowledgments. This research was supported by Basic Science Research Program through the National Research Foundation of Korea (NRF) funded by the Ministry of Education, Science and Technology (NRF-2017R1D1B03035040).

References

1. Rahman, M., Rahman, M., Carbunar, B., Chau, D.H.: FairPlay: fraud and malware detection in Google Play. In: Proceedings of the 2016 SIAM International Conference on Data Mining. Society for Industrial and Applied Mathematics (2016)

2. Verma, S., Muttoo, S.K., Pal, S.K.: MDroid: android based malware detection using MCM classifier. Int. J. Eng. Appl. Sci. Technol. **1**(8), 206–215 (2016)
3. Ham, Y.J., Moon, D.Y., Lee, H.W., Lim, J.D., Kim, J.N.: Android mobile application system call event pattern analysis for determination of malicious attack. Int. J. Secur. Its Appl. (IJSIA) **8**(1), 231–246 (2014)
4. Enck, W., Ongtang, M., McDaniel, P.: Understanding android security. IEEE Secur. Priv. Mag. **7**(1), 50–57 (2009)
5. Shabtai, A., Kanonov, U., Elovici, Y., Glezer, C., Weiss, Y.: Andromaly: a behavioral malware detection framework for android devices. J. Intell. Inf. Syst. **38**(1), 161–190 (2012)
6. Android-studio app-signing Homepage. https://developer.android.com/studio/publish/app-signing.html?hl = ko. Accessed 26 Feb 2018

Implementation of Security Mechanism in IIoT Systems

Minjeong Shin[1], Jihyeon Woo[1], Ibrahima Wane[1], Sungun Kim[1](✉), and Heung-Sik Yu[2]

[1] Department of Information and Communication Engineering, Pukyong National University, Busan, Korea
kimsu@pknu.ac.kr
[2] Unomic Technology, Busan, Korea

Abstract. The Fourth Industrial Revolution is the paradigm that realizes high-quality intelligent real-time services by combining super intelligent computing technologies associated with super connectable information and communication technologies for various industries. However, the implementation of real-time IoT services satisfying these needs is not possible with just applying existing Cloud technology. So, in this paper, we propose a network framework based on Cloud and Fog computing for the implementation of super connectivity network in IIoT systems for smart factory automation. Also in order to guarantee the confidentiality and integrity of the data to be processed between Fog and Cloud, an efficient security mechanism is needed to be designed and applied. The contribution of this research is also to provide efficient security mechanisms for the proposed service architecture.

Keywords: Industrial Internet of Things (IIoT) · Cloud & Fog computing
Network security · Lightweight Encryption Algorithm (LEA)
Message authentication tag · Diffie-Hellman key generation & exchange method

1 Introduction

In order to gather and analyze the enormous data generated in the infrastructure of IIoT (Industrial Internet of Things), a well-connected wide area network is essential. In this research, we propose a super-connected network framework based on Fog and Cloud computing. And more we suggest an implementation of security mechanisms which responds to the security threats existing in Fog and Cloud communications.

2 Target Network and IIoT System

2.1 Cloud Computing

Cloud computing technology is an on-demand computer environment that leases expensive storage and processing systems over the Internet [1]. By storing information in an external server, we can achieve the effect of system utilization, accessibility and scalability efficiently over the Internet. However, due to the huge bandwidth

S. O. Hwang et al. (Eds.): ICGHIT 2018, LNEE 502, pp. 183–187, 2019.
https://doi.org/10.1007/978-981-13-0311-1_30

consumption and latency, it is difficult to meet the targeted aim if we just adapt the cloud computing technology only. Therefore we need an alternative solution known as Fog computing, that extends the ability of super connectivity to the networks associated with IIoT systems for smart factories [2, 3].

2.2 Fog Computing

The Fog computing technology places a local server near the IIoT devices to realize the targeted real-time services effectively. So the sensitive real-time data is processed in the local server and the data that needs to be utilized after filtering is stored in a remote server [3]. This concept complements the limits of Cloud computing technology.

2.3 Cloud and Fog Computing for IIoT Systems

Generally, with Fog computing technology only, there is a limitation in the implementation of super intelligent services. However properly combining Cloud and Fog computing technologies can recover this limitation. The proposed network framework is illustrated in Fig. 1.

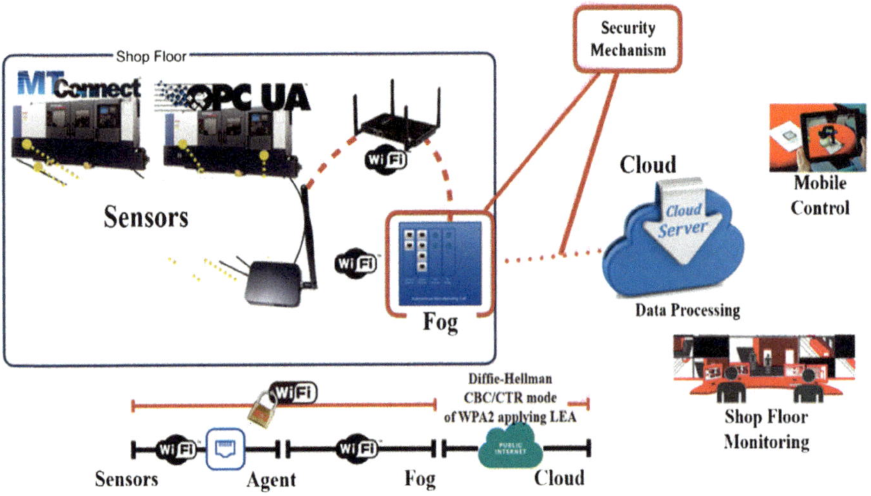

Fig. 1. The proposed network framework for real-time IIoT services

2.4 Security Problems

Table 1 summarizes possible security threats in the network architecture based on Fog and Cloud computing. In general, we study a security framework which guarantees data integrity and confidentiality in the aspect of the network scope summarized in Table 1 [4]. As far as the network concept is concerned, the security treats and their characteristics are explained in Table 1. Our security mechanisms are focused on these categories for providing an efficient method to IIoT systems for smart factory.

Table 1. Threats related to Cloud and Fog computing

Categories	Threats	Characteristics
Network related	Malicious intrusion	Concerns of malicious user or service intrusion such as DDos, Malware, or etc. [5]
	Routing hindrance	Routing hindrances such as Routing Information modulation and counterfeiting, sending error message, or etc. [5]
	Network failure	Service interruption due to Network hindrance
Cryptography algorithm related	Encryption algorithm	Threats for data integrity or restriction on use of database utilization services due to incorrect algorithm implementation [5–7]
	Key generation	Threats for confidentiality and integrity of key due to third party involvement during generating the key [5–7]
	Integrity	Requires authentication mechanism against message modulation by attackers [5–7]

2.5 Security Mechanism

In order to guarantee the stability of the transmitted data in the networks based on Fog and Cloud, there are two security mechanisms. One is the method using encryption and decryption technology. And the other is applying message authentication tag. In the encryption and decryption approach, it is important to generate secured secret key management and to use a secure encryption and decryption algorithm. In general, if we apply just this approach, it will not be enough to cope with the data forgery and modulation, so we need to use the message authentication tag.

Message authentication is a necessary mechanism for the proposed network framework because it can protect against data tampering or counterfeiting threats from third parties.

3 Security Mechanism Implementation

In this paper, we implement a security mechanism suitable for the proposed system in three parts: encryption and decryption, integrity and authentication, and key generation and exchange. All the algorithms are developed based on the block cipher encryption approaches (proposed for WPA-2) for the proposed network framework given in Fig. 1. Actually they utilize AES algorithms in the CTR (Counter) mode of WPA-2 to perform encryption and decryption and to apply the CBC (Cipher block chaining) mode of WPA-2 to generate the message authentication tag.

3.1 Encryption and Decryption Mechanism

We choose the LEA (Lightweight Encryption Algorithm) that is a lightweight cryptographic technology developed by the National Institute of Security in Korea [8].

Here we apply the CTR mode of WPA-2 to perform the encryption and decryption procedure by applying the LEA algorithm instead of the AES algorithm [9].

3.2 Integrity Authentication Tag

For the integrity authentication, with applying the LEA algorithm, we propose the CBC mode of WPA-2 to generate the message authentication tag [9].

3.3 Key Generation and Exchange

In order to avoid the complexity of key generation and exchange between sender and receiver, we investigate on adapting the Diffie-Hellman key generation and exchange mechanism in each part automatically [10].

3.4 Implementation of Security Mechanism

As illustrated in Fig. 1, the implementation of security mechanisms is given in Fig. 2.

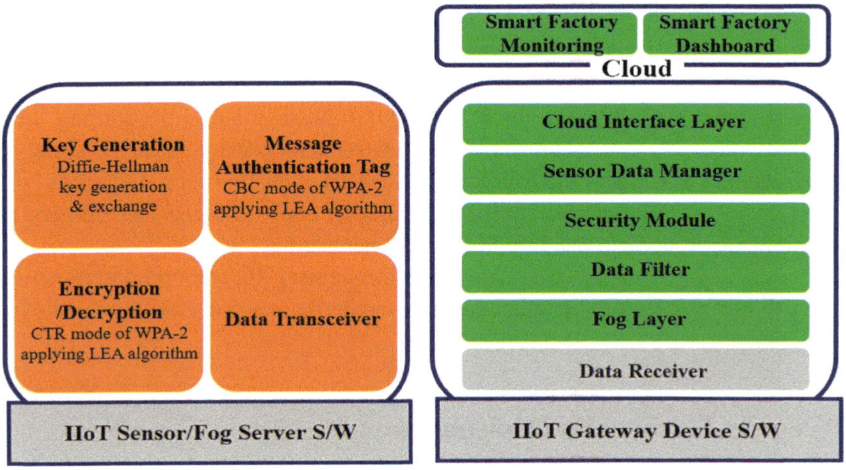

Fig. 2. The proposed security mechanisms between Fog and Cloud

First, the message authentication tag is generated by the method explained in Sect. 3.2. In this process, the key generation and exchange is accomplished automatically by the method proposed in Sect. 3.3. Finally the message to be sent encapsulated with the message authentication tag is encrypted by the method given in Sect. 3.1.

4 Conclusion

We are implementing the proposed security mechanisms. In the future, we will apply it to some real systems on site and evaluate the performance of the proposed mechanisms compared with other methods.

Acknowledgment. This work was supported by the Korea Technology and Information Promotion Agency for SMEs (TIPA) grant funded by the Korea Government (Ministry of SMEs and Startups) (2017, No. C0566220).

References

1. Zhang, Q., Cheng, L., Boutaba, R.: Cloud computing: state-of-the-art and research challenges. J. Internet Serv. Appl. **1**(1), 7–18 (2010)
2. Diaz, M., Martin, C., Rubio, B.: State-of-the-art, challenges, and open issues in the integration of internet of things and cloud computing. J. Netw. Comput. Appl. **67**, 99–117 (2016)
3. Bonomi, F., Milito, R., Zhu, J., Addepalli, S.: Fog computing and its role in the internet of things. In: Proceedings of the First Edition of the MCC Workshop on Mobile Cloud Computing, pp. 13–16 (2012)
4. Shin, M., Kim, S.: A study on the security framework for IoT services based on cloud and fog computing. J. Korea Multimed. Soc. **20**(12), 101–112 (2017)
5. Vitti, P.A.F., dos Santos, D.R., Westphall, C.B., Westphall, C.M., Vieir, K.M.M.: Current issues in cloud computing security and management. Secuware **2014**, 36–42 (2014)
6. Sarkar, S., Bharadwaj, V.K., Priya, G.: Security issues and challenges in cloud computing. Int. Res. J. Eng. Technol. **3**(10), 1143–1146 (2016)
7. Arora, R., Parashar, A.: Secure user data in cloud computing using encryption algorithms. Int. J. Eng. Res. Appl. **3**(4), 1922–1926 (2013)
8. Park, J.: 128bit block cipher LEA. TTA J. **157**, 64–69 (2015)
9. Dworkin, M.: Recommendation for block cipher modes of operation. Methods and techniques (No. NIST-SP-800-38A). National Institute of Standards and Technology (2001)
10. Rescorla, E.: Diffie-Hellman Key Agreement Method. IETF RFC2631 (1999)

An Efficient Searchable Encryption Scheme in the Multi-user Environment

In Tae Kim[1], Tran Hai Quan[2], Ly Vu Duc[2], Trong Kha Nguyen[2], and Seong Oun Hwang[1(✉)]

[1] Department of Computer and Information Communication Engineering, Hongik University, Jochiwon, Sejong 30016, Korea
sohwang@hongik.ac.kr

[2] Department of Electronics and Computer Engineering, Hongik University, Jochiwon, Sejong 30016, Korea

Abstract. Searchable encryption scheme is a cryptographic one that allows search of specific information in an encrypted content without decryption. There have been two main issues with searchable encryption: search time and extension from single reader to multi-reader setting. In this paper, we propose an efficient searchable symmetric encryption scheme by solving these issues.

Keywords: Searchable encryption
Searchable symmetric encryption · Cryptography

1 Introduction

Searchable encryption scheme is a cryptographic one that allows search of specific information in an encrypted content without decryption. Suppose the data (files, documents, emails, etc.) of a user is stored on untrusted server/gateway. To reduce security and privacy risks, the data should be encrypted.

When the user wants to retry the data which contain the keyword w, the best solution is *Searchable Encryption (SE)*.

SE enables the server/gateway to search for the data containing the keyword without decrypting the encrypted data, and thus, with the minimal damage to data confidentiality.

SE schemes work on the client/server model, where clients request servers to store or search encrypted data on behalf of them. In each case, the clients become the writers or readers, respectively. As a result, SE schemes can be divided into the following four SE architecture:

- single writer/single reader (S/S)
- multiple writer/single reader (M/S)
- single writer/multiple reader (S/M)
- multiple writer/multiple reader (M/M)

S. O. Hwang et al. (Eds.): ICGHIT 2018, LNEE 502, pp. 188–192, 2019.
https://doi.org/10.1007/978-981-13-0311-1_31

SE have been proposed in two main directions: searchable symmetric encryption (SSE) [1,2] and public key encryption with keyword search (PEKS).

In SSE, to search data from a storage server, a client can generate search tokens of encrypted data as queries to the storage server. One approach to provisioning symmetric encryption with search capabilities is the so-called secure index. An index are considered to be a data structure that uses to store document collections for supporting efficient keyword search. I.e., given a keyword, the index is used to return pointers to the documents that contain it (Table 1).

Next, we explain the basic concept of PEKS [3–5]. Suppose that an email server stores many emails encrypted using the corresponding receiver's public key, which contains a number of keywords. A receiver produces a trapdoor using his secret key and a keyword w. In turn, a receiver can retrieve emails containing a specific keyword w by giving a corresponding trapdoor to the email server without information leakage.

Table 1. Comparison between PEKS and SSE.

PEKS (public-key)	SSE (secret-key)
Support of M/S, M/M	Support of S/S (Some supports S/M)
Worse performance: no index, no keyword	Better performance: index, keyword
Application: email routing system, encrypted file sharing system	Application: data storage

2 Related Work

Symmetric key primitives allow a single user that has secret key which use for encryption and decryption to read and write data (S/S). The first S/S scheme is proposed by Song et al. [1,2]. Because it uses symmetric key cryptography, only the secret key holder can create searchable ciphertexts and trapdoors.

By contrast, in a public key encryption (PKE) scheme, a public key and a secret key are used as follows, respectively. The public key can be used to encrypt a message by anyone, and the secret key is used to allow the holder to decrypt the encrypted message by using the paired public key. Thus, PKE can allows many users to write data using the public key. However, only the holder of the paired secret key can perform search process. Therefore, searchable encryption based on PEK, i.e., PKES, can support M/S architecture. The first M/S scheme is proposed by Boneh et al. [4].

3 Issues with Searchable Encryption

3.1 Search Time

To improve the efficiency of search process, an index extracted from the unencrypted data is used in databases as a common tool. As a result, by introducing an index to searchable encryption, the search complexity can be significantly decreased and thus the search performance of a scheme is increased. There are two types of index:

- Forward Index.

  This type of index will reduces the search time to the number of documents, i.e. search time is linear in the number of documents. It is not suitable for large-scale databases (Fig. 1).
- Inverted Index.

  This type of index is more optimal than forward index above. This reduces the search time to the number of documents that contain the keyword. However, this technique is more suitable for a static database than a dynamic one (Fig. 2).

The issue is how to build a scheme based on inverted index.

document id	keywords
1	w_2, w_5, w_7
2	w_1, w_2, w_4, w_6, w_8
...	...
n	w_2, w_5, w_6

Fig. 1. Forward index.

keyword	document ids
w_1	$2, 3, 9$
w_2	$1, 2, 6, 7, n$
...	...
w_m	$1, 3, 8$

Fig. 2. Inverted index.

3.2 Extension from Single Reader (*/S) to Multi-reader (*/M) Setting

Some schemes extending */S to */M will need the key distribution conducted by key sharing (through secure channel) or proxy re-encryption. The issue is how to extend from single reader (*/S) setting to multi-reader (*/M) without using key distribution.

4 Proposed Searchable Encryption Scheme

In this section, we propose our scheme to solve two issues in SE as discussed above. The construction consists of six algorithms: *Setup, BuildIndex, AddUser, Search1, Search2, RevokeUser.*

- *Setup* $(1^k, p)$: Generate $K \leftarrow \{0,1\}^k$. $\alpha \in Z_p^*$. h_1, h_2, h_3 are hash functions.
- *BuildIndex:* Build an array A and look-up table T.
  1. From a document collection $D = \{D_1, D_2, \ldots, D_d\}$ and a set of distinct keywords $W = \{w_1, w_2, \ldots, w_n\}$, build $D(w_i) = \{id(D_{i,j})\}$ is a set of identifiers of documents containing keyword w_i, j is the jth identifier in $D(w_i)$.
  2. For each keyword w_i:
     - Set node $N_{i,j}$ in linked list L_i with $1 \leq j \leq |D(w_i)|$: $N_{i,j} = \langle id(D_i, j) || pointer(j+1) \rangle$ with $pointer(j+1)$ is the pointer indicating the next node $N_{i,j+1}$.
     - For the last node in L_i, set the address of the next node is null.
  3. Let $d' = \sum_{w_i \in W} |D(w_i)|$. If $d' < d$,[1] set remaining entries of A to random values.
  4. For each keyword w_i, set $\gamma = addr(N_{i,1}) \oplus h_1(w_i)$; Set $T[h_2(e(g^\alpha, h_3(w_i)))] = T[\gamma(w_i)]$.
  5. Output $I = (A, T)$ and the encrypted document collection $[D]_K$.
- *AddUser*: U is the owner of document collection D. U adds a user U_n to a group that is allowed to search on encrypted data of U as follows:
  1. Generate r_n randomly from Z_p^* and send (K, r_n) to U_n.
  2. Compute $g_n = g^{\alpha/r_n}$ and send $(id(U_n), g_n)$ with $id(U_n)$ is the identifier of U_n to Server. Server adds this entry into *AllowList*.
- *Search1:* U searches for keyword w on his own encrypted data as follows:
  1. Send $(h_1(w), \gamma'(w) = h_2(e(g^\alpha, h_3(w)))$ to Server.
  2. If $T[\gamma'(w)] = T[\gamma(w)]$, from $addr(N_{i,1})$, Server obtains all nodes in L_i and returns the list of document identifiers to U. Else, return $\perp$ to U.
- *Search2:* U_n searches for keyword w on U's encrypted data as follows:
  1. Send $(id(U_n), h_3(w)^{r_n})$ to Server.
  2. Server checks in *AllowList*. If there is no entry for $id(U_n)$, return $\perp$ to U_n. Else, obtain g_n and compute $h_2(e(g_n, h_3(w)^{r_n})) = \gamma(w)$.
  3. If $T[\gamma(w)] = T[\gamma]$, from $addr(N_{i,1})$, Server obtains all nodes in L_i and returns the list of document identifiers to U_n. Else, return $\perp$ to U.
- *RevokeUser:* U sends delete request to Server to delete an enrty of U_n from *AllowList*.

[1] d is the size of document collection and A, $d' = d$ in the case all keyword w_i appears only once in a document in D.

5 Conclusion and Future Work

Searchable encryption scheme is considered to be one of widely used cryptographic ones in near future because it allows search of specific information without losing confidentiality of data. However, the existing schemes increases search time by using forward index and work under restricted environment like single reader setting. To solve these issues, we propose an efficient searchable symmetric encryption scheme by using inverted index and extending single reader to multi-reader setting.

Most schemes including the proposed scheme support search on exact keyword, that is, the server returns the set of documents containing keyword or nothing. How about the case if keyword is not given exactly? This case is very likely to happen in the real world. As future work, we plan to develop searchable encryption schemes which support similar keywords or diverse combinations of keywords.

Acknowledgment. This work was supported by the National Research Foundation of Korea(NRF) grant funded by the Korea government(MSIP) (No. 2017R1A2B4001801).

References

1. Curtmola, R., Garay, J., Kamara, S., Ostrovsky, R.: Searchable symmetric encryption: improved definitions and efficient constructions. J. Comput. Secur. **19**(5), 895–934 (2011)
2. Song, D. X., Wagner, D., Perrig, A.: Practical techniques for searches on encrypted data. In: Proceedings of the 2000 IEEE Symposium on Security and Privacy, pp. 44–55. IEEE Computer Society, IEEE (2000)
3. Hwang, Y.H., Lee, P.J.: Public key encryption with conjunctive keyword search and its extension to a multi-user system. In: International Conference on Pairing-Based Cryptography. LNCS, vol. 4575, pp. 2–22. Springer, Heidelberg (2007)
4. Boneh, D., Di Crescenzo, G., Ostrovsky, R., Persiano, G.: Public key encryption with keyword search. In: International Conference on the Theory and Applications of Cryptographic Techniques. LNCS, vol. 3027, pp. 506–522. Springer, Heidelberg (2004)
5. Dong, C., Russello, G., Dulay, N.: Shared and searchable encrypted data for untrusted servers. IFIP Annual Conference on Data and Applications Security and Privacy. LNCS, vol. 5094, pp. 127–143. Springer, Heidelberg (2008)

Verification Test of Cloud PaaS (PaaS-TA) on Microserver

Byung-Kwon Jung[1(✉)], Zi-Ho Shin[2], Hyuk-je Kwon[1], Jin-Suk Ma[1], Hag-Young Kim[1], and Dong-jae Kang[1]

[1] Cloud Computing Research Group, Electronics and Telecommunications Research Institute (ETRI), Gajeong-dong, Yuseong-gu, Daejeon, Korea
bkjung@etri.re.kr

[2] Department of Computer S/W, University of Science and Technology Daejeon, Daejeon, Korea

Abstract. In this paper, we fabricated a modular microserver using low power chip Intel ATOM CPU or ARM based XGEN2 CPU. Cloud Platform-as-a-Service (PaaS) is a promoting software technology in commercial cloud application. PaaS-TA provides a web-based platform management user portal platform, including cloud-based development operating environment management and PaaS installation automation technology, which is basically required for PaaS operation. We installed PaaS-TA, a Cloud Platform (PaaS), on a highly integrated cluster server, and verified that various PaaS services operate normally. As a result, we confirmed the versatility, scalability, and interoperability of the developed microserver and open cloud platform, PaaS-TA.

Keywords: Microserver · Openstack · IaaS · PaaS · PaaS-TA

1 Introduction

Currently we are living in an era of massive data explosion. With the spread of digital devices such as smartphones and laptops, mobile users easily have access to the Internet anywhere in the world so that they can share photos and other information. As the amount of worldwide data is increasing by 40% each year, the innovation of computing architecture design is indispensable to compute such vast amounts of data. And in order to receive cloud services, a flexible and cost-effective cloud platform should be provided. In order to realize various functions of cloud computing, performance improvement of server system has been demanded, and leading processors and server manufacturers such as Intel, ARM, HPE, and Supermicro have satisfied this demand through high integration of processors.

In this paper, we confirmed the versatility, scalability, and interoperability of the microserver developed by our research group. The procedure was done by verifying that the PaaS-TA environment, which is a PaaS platform, is installed and operating normally after constructing OpenStack in the micro server. Subsequent paragraphs, however, are indented.

S. O. Hwang et al. (Eds.): ICGHIT 2018, LNEE 502, pp. 193–198, 2019.
https://doi.org/10.1007/978-981-13-0311-1_32

2 Microserver System

2.1 Microserver

The microserver provides a highly integrated and resource-efficient environment in a cluster structure by making Intel low-power ATOM CPU and APM's ARM-based XGEN2 CPU as a modular server. Recently, servers based on ARM and ATOM architecture-based processors, which were used in the mobile and embedded markets, are beginning to emerge. This has been a starting point for research on microservers of domestic and foreign leading companies [1]. Figure 1 shows the microserver developed by this research group.

Fig. 1. Microserver of ETRI

Microservers have been developed with the goal of achieving performance comparable to existing servers through a highly integrated mobile processor as well as the low power system implementation at the same time.

2.2 Openstack

OpenStack is an open source project for cloud computing in the form of Infra-as-a-Service (IaaS). OpenStack consists of several subprojects to control the available resources such as processing, storage space and networking. Users can use the web-based Dashboard, command line interface (CLI) or application program interface can be managed. In addition, development is being done in an open source format, with more than 500 companies participating in the OpenStack development project and distributing new versions every six months. The version used in this paper is the OpenStack Mitaka version [2].

2.3 PaaS-TA

PaaS-TA is an open cloud platform for e-government created by Korea Information Science Promotion Agency (NIA). The first version was released in June of 2016 and the services are continuously developed and supported by opening 2.0 version, released in February this year [3]. PaaS-TA 1.0 has a management function of cloud-based development and operation environment, basically required for Platform-as-a-Service (PaaS) operation. It also offers a domestic SW and e-government framework. PaaS-TA 2.0 provides additional features such as Web-based platform management user portal platform-based automatic inter-tier monitoring extension, and metering for usage-based billing support. Currently, PaaS-TA 2.0 supports vSphere, AWS, and OpenStack.

Nowadays, the platform market is growing at a high rate because it is necessary to develop a platform that can reflect various conditions related to consumers' needs. Especially, in the case of enterprises, it is necessary to operate efficiently according to the development of technology and new desires of consumers. By using these platforms, it is easy to respond effectively when it comes to supply of raw materials and parts and utilization of infrastructure.

3 PaaS-TA Verification Environment and Results

3.1 Openstack Environment

The OpenStack used in this paper consists of a total of 9 nodes: one controller, seven compute nodes, and one block storage node [4]. Figure 2 is a list of OpenStack compute services before installing the PaaS-TA. Figure 3 illustrates the testbed network architecture. The network for PaaS is composed of one external network, two internal networks, and three routers.

Id	Binary	Host	Zone	Status	State	Updated At
5	nova-consoleauth	controller	internal	enabled	up	2017-09-07T11:31:55.000000
6	nova-scheduler	controller	internal	enabled	up	2017-09-07T11:31:55.000000
7	nova-conductor	controller	internal	enabled	up	2017-09-07T11:31:54.000000
15	nova-compute	compute1	resource	enabled	up	2017-09-07T11:31:57.000000
16	nova-compute	compute2	resource	enabled	up	2017-09-07T11:31:52.000000
17	nova-compute	compute3	resource	enabled	up	2017-09-07T11:31:57.000000
18	nova-compute	compute6	resource	enabled	up	2017-09-07T11:31:52.000000
19	nova-compute	compute4	resource	enabled	up	2017-09-07T11:31:57.000000
20	nova-compute	compute7	resource	enabled	up	2017-09-07T11:31:50.000000
21	nova-compute	compute5	resource	enabled	up	2017-09-07T11:31:54.000000

Fig. 2. Openstack compute Service list

3.2 PaaS-TA Verification Environment

In order to verify that all functions of PaaS-TA operate properly in the environment of OpenStack, 61 functional items (unit test, integration test) and 11 nonfunctional test items (scalability, safety, performance, security) were tested to see if all the target functions passed. Table 1 presents the classification of the PaaS-TA tests and the number of items (Table 2) (Fig. 4).

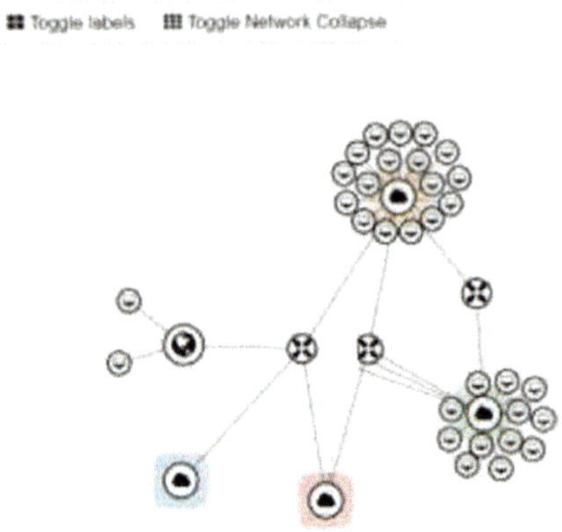

Fig. 3. Network structure of Testbed

Table 1. Range of PaaS-TA test

Classification		Division	Number of test items
Function test	Unit	Cloud infra	2
		Bosh micro	2
		Release	1
		Stemcell	1
		Complied package	1
		Deployment	3
		Status	1
		VM	1
		Bosh account management	2
		Org management	7
		Space management	4
		User management	4
		Domain management build package Management	3
		App management	4
	Integration	Install	4
		Distribute	5
		Operation management	2
		Security	2
Non-function test		Scalability	2
		Stability	3
		Performance	2
		Security	3

Table 2. Function integration test item and result

		Test case	Input data	Test result
Total test	Install and verification	Q: Is the inception properly installed? (Bosh CLI) A: Create IaaS credential and key pair A: Create a dedicated inception VM	IaaS credential	Pass
		Q: Is Bosh installed normally? A: Find the latest Bosh Stemcell in the public store and download it A: Install Micro Bosh with Bosh deploy command	Release, Stem cell, Manifest	Pass
		Q: Is CloudFoundry/CF CLI installed normally? A: After downloading the latest CF release and Stemcell, create a release to be deployed A: Upload to Bosh release and Stemcell	CF credential	Pass

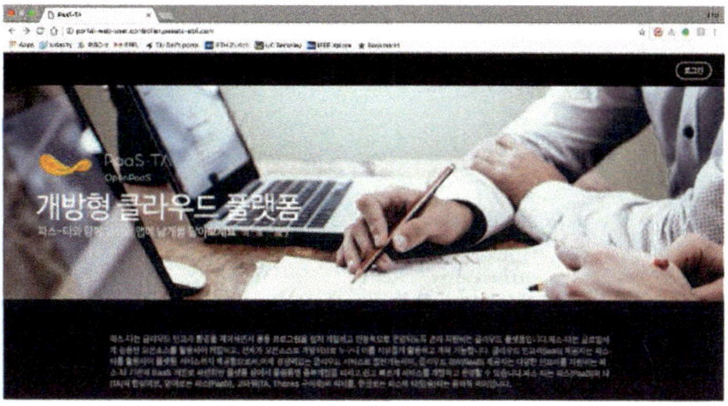

Fig. 4. Builted PaaS-TA's browser

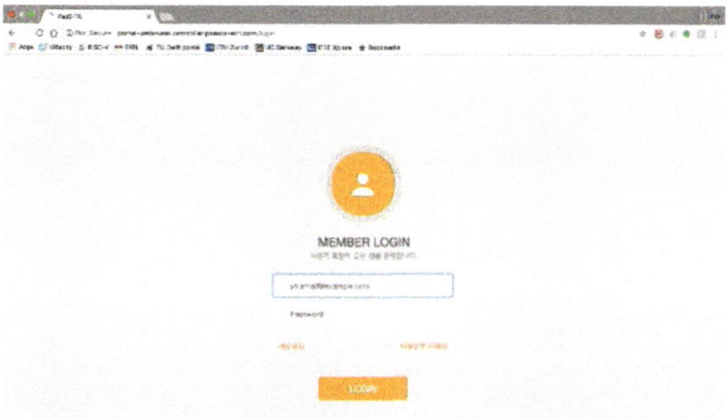

Fig. 5. Builted PaaS-TA environment

4 Conclusion and Future Works

In this paper, PaaS-TA environment is established as an open cloud platform for e-government in the microserver developed by this research group. We verify versatility, scalability, and interoperability of the developed microserver by showing that various PaaS services operate normally in the constructed environment (Fig. 5).

In the future, for scalability and general usability, it is necessary to continuously improve the product quality in the aspect of hardware and software of the microserver in order to spread the product in the domestic server market.

References

1. Kwon, W.: Trend technology of microserver. Electr. Telecommun. Trend Anal. ETRI **148**, 49–58 (2014)
2. Wikipedia Homepage. https://ko.wikipedia.org/wiki/%EC%98%A4%ED%94%88%EC%8A%A4%ED%83%9D
3. PaaS-TA Homepage. https://paas-ta.kr/intro/component
4. Shin, Z.: Multi-node open stack installation and operation of micro server system. In: Summer Conference of Electronics Engineering Society of Korea, pp. 1395–1398, June 2017

Smart City and Business Model with a Focus on Platform and Circular Economy

Junghee Han(✉) and Huy-Doo Jin

Graduate School of Smart City Science Management,
Hongik University, Seoul, Korea
Hjh0037@hongik.ac.kr

Abstract. City development is related with the evolution of social, economic, technological process. City itself has been paid attention to job creations since emerging the Smart cities. Smart city seems to be the convergent point for all process. Business model must be adaptive over times in response to changing the markets, technologies, environmental, and economic circumstance. For companies, being increasingly scarce in resources business model is becoming business constraints. Under the circular economy the principles of a new approach to business modelling is necessary. Limited exploitation of the resource in earth planet forces firms to change their business model. The aim of this paper is to suggest pertinent business model in circular economy. To fulfill it, a qualitative methodology is used. Cost and convenience for both firms and consumers are the key critical factors to be channeled in the new business model for innovation to achieve the greening and efficiency objectives.

Keywords: Smartcity · Business model · Circular economy · Platform

1 Introduction

This study was initially motivated in order to provide new business model as advent of Smart cities. In association with economy or jobs Smart City is used to describe a city with a "smart" industry. That implies especially industries in the fields of information and communication technologies (ICT) as well as other industries implying ICT in their production processes.

This paper aims to provide pertinent highlights on the business model designed in circular economy (CE) at Smart city environmental condition for entrepreneurs. Namely, this paper seeks to a new conceptual framework for business models from a research question perspective; which business models do emerge in circular economy area? Business models evolve over times (Teece 2010; Chesbrough 2002; Zott et al. 2011). The advent of circular economy might be caused by the concept of sharing economy and environmental and sustainability technologies. Also, relationship between industry and environment is important for business performance. Since planet earth's resources are limited, ceaseless exponentially economic growth cannot be met without serious consequences for health and environment (Meadows et al. 1972).

In this paper, the concept of circular economy (CE), sustainable development is development that meets the needs of the present without compromising the ability of

S. O. Hwang et al. (Eds.): ICGHIT 2018, LNEE 502, pp. 199–203, 2019.
https://doi.org/10.1007/978-981-13-0311-1_33

future generations to meet their own needs. Circular business models may obtain a greater competitive edge in the years to come because they create more value from each unit of resource than the traditional linear production and consumption model. It is needless to say that innovation is related to the firm growth. Innovation, environment and economics are closely inter-related. Finally, the business model for innovation is proposed as a way forward towards implementing the comprehensive CE approach in the industry sector. This paper offers five business models: green-loop model, pass-loop model, having-loop model, being-loop model, and hybrid-loop model.

This paper is organized in five sections where Sect. 2 presents the research methodology in detail and state of current research on CE. Once the circular economy notion is clarified, in the Sect. 2.1 researches on business model innovation are discussed. Section 3 presents business model typology in circular economy. Finally, in Sect. 4, conclusions are drawn from the performed research.

2 Literature Review and Methodology

This paper defines smart city as enabling platform because the general concept of Information, communication and technology (ICT) underpins the idea of smart technology. In other words, smart technology overlaps with and is a key part of ICT. As the World Bank puts it, ICT is comprised of "hardware, software, networks, and media for the collection, storage, processing, transmission and presentation of information (voice, data, text, images), as well as related services. The author considers smart technology in the context of smart cities as espousing most elements of broader ICT definitions—hardware, software, and services. Put another way, smart technology serves as the connective tissue that binds the individual components, including home, office, mobile phone, and car, on a single, access-agnostic technology platform. A strong smart technology framework goes beyond point-solutions and brings seamless connectivity through the adoption of high-speed broadband connections, the Internet of Things (IoT), the provision of free Wi-Fi, 4G, and, in the future, 5G technology. Ultimately, smart technology lays the foundation needed for vertical-specific applications and technologies to function. A significant function of smart technology is thus related to connectivity and communications services that serve as the foundation for a plethora of citizen-facing smart city applications. The role of a platform is to mediate between various stakeholder constituencies by coordinating interactions between two or more members from distinct groups of stakeholders (Evans 2003; Rochet and Tirole 2003; Jullien 2005; Cortade 2006). Jacobides et al. (2006) emphasized that a platform creates value and is an important structure of the industry architecture in the ICT sector.

2.1 Business Model in Circular Economy

A business model is the way how a business creates and delivers value to customer (Teece 2010). In short, a business model could be defined as the manner in which an enterprise creates and delivers value to customers, and then converts the payments that are received into profit (Björkdahl 2009; Chesbrough 2007; Zott and Amitt 2007; 2010; Zott et al. 2011). Business models are defined in terms of a firm's economic model

(Morris et al. 2005). The business model concerns the logic of how profit is generated by the firm. Steward and Zhao (2000) assert that the business model is a statement of how a firm will both make money and sustain its profit stream over time. Furthermore, the business model can be defined as an architectural configuration of the firm's survival. Slywotsky (1996) noted that a business model consists of the totality of how a company selects its customers, defines and differentiates its offerings, defines the task it will perform itself and those it will outsource, configures its resources, goes to market, creates utility for customers, and captures profits.

A business model is related to a number of other managerial concepts, and it captures the key components of a business plan. The business model components consist of price, product, distribution, organizational characteristics, and market strategy (Horowitz 1996; Timmers 1998; Dubosson-Torbay et al. 2001; Hamel 2001). Business model must be evaluated against the current state of the business ecosystem, and against how it might evolve (Teece 2015). No matter the industry sectors where a firm belongs to, there are criteria that enable one to determine whether or not one has designed a good business model.

Business models yield value propositions that are compelling to customers, achieve advantages in terms of the cost risk structures, and enable significant value to be captured by the business (Han and Cho 2015). Prior studies are referenced in this study to define the business model of a company as a simplified representation of its business logic. The business model is defined in the ways of how a firm survives in terms of delivering and capturing values. Recently business model research area is the sustainability economy. Bolton and Hannon (2016) suggested that a systems based approach to the analysis of business models as embedded in their socio-technical contexts can offer new insights into the dynamics and governance of sustainability transitions. Business model can flow the value, sharing profits between two strategic partners, manufacturer and retailer to maximize an individual profits as well as total profits concurrently in reverse supply chain (Yoon and Jeong 2016) (Figs. 1 and 2).

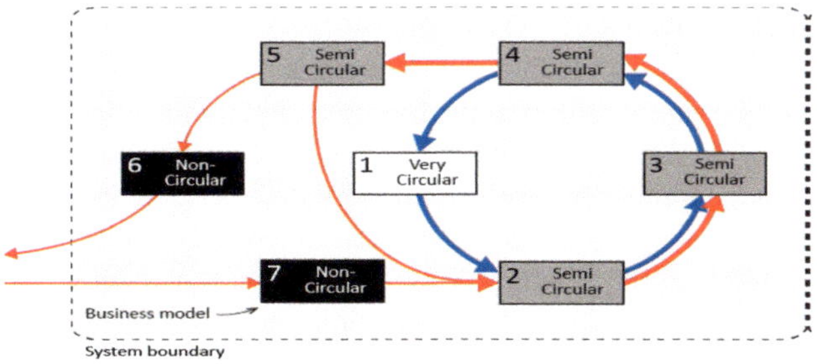

Fig. 1. Circular economy concept

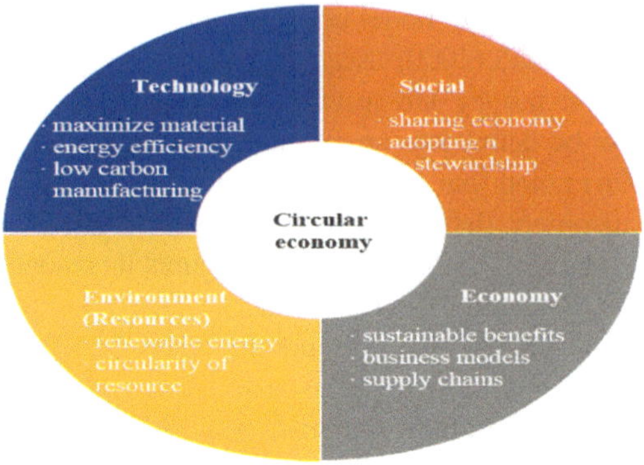

Fig. 2. Circular economy business model

3 Business Model Typology in Circular Economy

As mentioned above, our current economy's feature (i.e. production, consumption) patterns a linear (take-make-use-dispose) approach to resources use. The transition from current linear ways of doing business to the circular economy approach will not be easy. Fully 100% linear business model and 100% circular business model do not exit since there are several limitations such as leakage of energy in the implementation level. In spite of it, from the findings, in contrast to today's linear model ('take-make-use-dispose'), a circular economy offers a development strategy that enables economic growth by capturing value in using resources efficiently through circular business models.

In fact, circular economy aiming to create added value in products for as long as possible and eliminate waste. The transition from a linear process to a circular process requires change in the use of materials and also a change in structures of ownership, business models and responsibilities. BMCEs offer instruments for translating products designed for re-use into attractive value propositions.

This perspective requires manufacturers to think differently about products and to take responsibility for products during the life cycle. Following questions are needed to be asked in order to find how the used products can be made valuable again: (1) what business models are appropriate for a specific product? (2) Can the product be offered as a service rather than be sold as a product? (3) Which optimization options are available for restorative (reuse, recycle, re-manufacturing, reduce) for production-chains? (4) How can value flows (social, natural and/or economic) be translated to allow for better charting of the benefits of the circular economy? And last but not necessarily least, (5) How can design to be manufactured and distributed?

In this paper, five business models are focused on: (1) green-loop model, (2) pass-loop model, (3) having-loop model, (4) being-loop model, and (5) hybrid-loop model. All models suggested have common principles (i.e. economic benefit, resource scarcity, environmental impact), but each model has own characteristics respectively.

Initially business model for a circular economy (BMCE) innovation was driven by mainly start-ups. In order to keeping a change for firms and customers, they should seek new business models and innovation under the circular economy. Most importantly, they choose the business model that is right for their business even though no one know right things for a firm to succeed in the upcoming circular economy. In adjusting or exploitation the new business models, firms make sure to have a business ecosystem including external enablers.

New business models, no matter what kinds of business model are chosen should contain that waste is eliminated or minimized, and monetized. And they should serve customers, and drive business and product development over time. Finally, new business model should be designed following and operating circular economy business principles creating a circular flow from making to production, taking–back, and profitable regeneration and reuse in the closed material loop.

4 Conclusion

This paper suggests five business models. In the beginning phase, while it takes a long time for them to become mainstream, it is conceivable that firms be adopted step by step to a minority of existing business. In the long run, this new approach can give rise to the evolution of new business model.

References

Abolhosseini, S., Heshmati, A.: The main support mechanisms to finance renewable energy development. Renew. Sustain. Energy Rev. **40**, 876–885 (2014)

Demil, B., Lecocq, X.: Business model evolution: in search of dynamic consistency. Long Range Plan. **43**, 227–246 (2010)

Lieder, M., Rashid, A.: Towards circular economy implementation: a comprehensive review in context of manufacturing industry. J. Clean. Prod. **115**, 36–51 (2016)

Shafer, S.M., Smith, H.J., Linder, J.C.: The power of business model. Bus. Horiz. **48**, 199–207 (2005)

Teece, D.J.: Business model, business strategy and innovation. Long Range Plan. **43**, 172–194 (2010)

Timmers, P.: Business models for electronic market. Electron. Mark. **8**(2), 3–8 (1998)

Troy, L.C., Hirunyawipada, T., Paswan, A.K.: Cross-functional integration and new product success: an empirical investigation of the findings. J. Mark. **72**, 132–146 (2008)

Van den Berg, M.R., Bakker, C.A.: A product design framework for a circular economy. In: Proceedings of the PLATE Conference, Nottingham, UK (2015)

Vermander, B.: A growth engine reinvents itself: towards a greener China? In: Greening Economic Growth: Towards a Global Strategy for Europe, p. 85 (2008)

Wicki, S.: Exploration of the business model - sustainability link in the case of the Westland greenhouse sector. (MSc Industrial Ecology), TU Delft/Leiden University (2013)

Yoon, S.W., Jeong, S.: Effects to implement the open-innovation coordinative strategies between manufacturer and retailer in reverse supply chain. J. Open Innov. Technol. Mark. Complex. **3** (2) (2016)

Yuan, Z.W., Bi, J., Moriguichi, B.: The circular ecology: a new development strategy in China. J. Ind. Ecol. **10**, 4–8 (2006)

The Application of IoT Technology in Monitoring Odors in Industrial Areas: The Case of Sasang Industrial Area in Busan

Ji-in Chang(✉)

Graduate School of Smart City Science Management, Hongik University, Sejong City, Korea
twomay_jiin@hongik.ac.kr

Abstract. Residents living close to Sasang Industrial Area (SIA) in Busan, South Korea, have long complained about unpleasant odors emitted by nearby factories. This paper shows how Busan municipal government, with the support of the central government, has effectively reduced residents' complaints by 80% in five years. A major contributing factor to this success was the use of IoT technology to monitor odors and pinpoint their sources. Sensors and CCTV cameras were installed throughout SIA to measure and record changes around the clock. These real time data and visual images were relayed to the Integrated Odors Control Center located within the Sasang District office building. The Control Center then shared the information with relevant government authorities and the public.

Keywords: Odor detection · Sensor · Integrated Odors Control Center Korea

1 Introduction

Residents living near industrial areas are often subject to unidentified odors emitted from nearby factories. Odorants are chemical substances unpleasant to the human sense of smell. They can also be deleterious to health and hinder the normal day-to-day activities for those affected. Air quality has been subject to increasing public attention, and people's awareness of the effects of these chemicals in air have sharpened the general sense of potential danger [1]. Odor nuisance is one of the major reasons for pollution complaints, as odors stimulate the human olfactory system at even low threshold levels [2].

The residents living in apartments near Sasang industrial area (SIA) in Busan have been complaining about odors for decades. In 1984, 10,000 students of all five schools around the SIA were so strongly affected by air pollution that they were unable to continue their classes, reporting nausea, respiratory difficulties, and eye inflammations [3]. The problem of air pollution was compounded by low lying terrain where air was unable to flow freely. For this reason, it earned the nickname of 'the windless place'. This meant that people living nearby were unable to open their windows even on hot summer days, with their laundry becoming discolored as they were hung out to dry.

S. O. Hwang et al. (Eds.): ICGHIT 2018, LNEE 502, pp. 204–209, 2019.
https://doi.org/10.1007/978-981-13-0311-1_34

Urban planning can be said to be responsible for the proximity of industrial land use to residential areas. The masterplan of Busan designated the location of SIA in 1965 so that factories which had been dispersed all over Busan area could access much needed land and infrastructure. The concentration of factories in this industrial area increased rapidly as the number of factories rose from 442 in 1975, to 1,582 in 1979, and to 2,506 in 1987 [3]. At this time, 39% of all factories located in Busan were concentrated in this area. Of all the factories located in SIA, 46.9% were categorized as mechanical industries, such as iron casting, steel etc., and 48.4% as chemical industries, such as electroplating, leather processing etc. Despite this heavy concentration of factories, the masterplan designated the areas east of SIA to be residential.

With the prevailing south-westerly winds, the odors and chemicals emitted from these factories traveled easily to the adjacent residential areas to the north and the east. Even though the number of factories and laborers began to decrease from the 1990s onwards, nearby residents have been directly and indirectly exposed to atmospheric pollution. The most serious case was in 2003 when an unidentified gas was released during the night, resulting in the hospitalization of eighteen people who lived nearby. With small and large incidences seriously affecting public safety, residents have become more vigilant and vociferous regarding air pollution and odors traveling to residential areas from SIA.

Since the 2010s, Busan Metropolitan Government has taken substantial steps to combat these unpleasant odors in terms of chemicals in the air. The results have been significant; complaints have decreased from 279 reported cases in 2011 to 123 cases in 2015 [4]. While the 50% decrease over four years were impressive enough, the number of citizens' complaints were further reduced by another 50% to 57 cases within one year between 2015 and 2016. This sharp reduction followed the opening of the Integrated Odors Control Center (IOCC) which applied IoT technology to collect, measure and publicly share atmospheric information in real time. Of course, the IOCC cannot take sole credit for this achievement. It can be attributed to a combination of multiple factors, such as Busan's environmental policy decisions, changes in demographics and land, as well as the application of 'smart technology'. Figure 1 shows the map of the SIA which is enclosed by mountains and located in a low lying terrain.

This paper delineates the measures taken with a focus on the recent application of IoT technology in monitoring and controlling odors in SIA.

2 Measures to Control Odors

Odor emission sources are composed of various odorous compounds. The main sources of odor emission in this area can be classified into four main categories as shown in Table 1 below [5, 6]. The odor intensity and concentration is measured by both stationary and mobile methods. The idea of installing instruments around the factory sites is to inform and contain the odors before they reach serious levels by sampling the ambient air through sensor chambers.

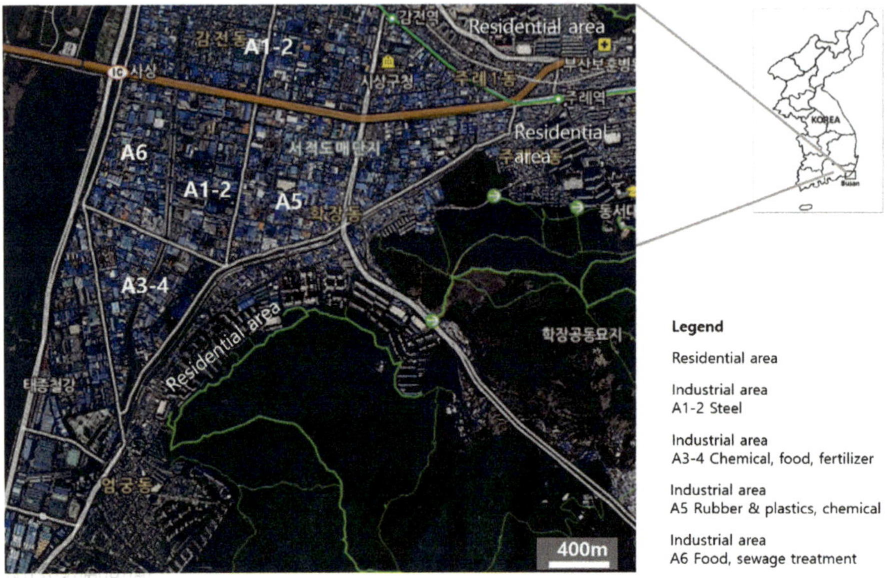

Fig. 1. Map of Sasang Industrial Area in Busan, Korea (Source: Modified map based on Lee et al. 2013)

Table 1. Classification of odor emission sources by odorants

Compound	Main odorants	Emission sources
Nitrogenous Compounds	Ammonia, Trimethyl Amine	Sewage disposal plant, rubber factory, rayon factory, landfill
Sulfur Compounds	Hydrogen Sulfide	Night-soil treatment plant
Aldehyde Compounds	Acetaldehyde	Metal coating factory, casting factory, off-set printing factory, car-coating factory
VOCs	Toluene, Xylene, Ethyl Acetate	Coating factory, furniture manufacturing factory, adhesive manufacturing factory, car repair shop

Source: Lee et al. (2013), 281 modified.

2.1 IoT Technology and the Integrated Odors Control Center

As the first of its kind in Busan, the IOCC started operation in September, 2015. Located in the room next to the Emergency Control Center within the Sasang District building, it collects and displays data from eleven stationary odor detection stations, eleven remote odor sampling stations, two high performance CCTVs, and three weather observation towers throughout the SIA. Twelve people take turns 24 h every day for 365 days a year to monitor and alert relevant authorities as soon as any complaints of odor is received.

In addition, Sasang District built an air quality modeling system and a web system specifically designed for smart phones at a cost of half a billion Won. The weather forecast is combined with odor emission information to provide forecasts based on real time measurements. Projections concerning the intensity of odor and the direction of the wind are thus possible (Fig. 2). Based on the data, the IOCC contacts relevant factories with requests to reduce emissions or to spray deodorizing agents so that odors are reduced before they reach a detectible level. These proactive measures have contributed to the sharp decline in complaints by residents living around SIA in 2016.

Fig. 2. Sasang IOCC central dashboard showing wind direction and speed, major sources of odor emission, CCTV view of factories. Source: Picture taken by author

The IOCC also operates a vehicle equipped with remote odor sampling sensor, an air quality sensor, and an electrical generator, among others (Fig. 3). This vehicle drives around the SIA and measures the intensity of the odors in various areas. Upon detection of excessive levels, the factory responsible is immediately notified and fined. At night, the vehicle remains near areas where odors are frequently emitted in the cover of darkness. Data is gathered and used later for unannounced detection visits.

All data collected by the sensors on site are transmitted to IOCC staff who have remote access to the measurement stations. It allows real time verification of data as well as monitoring and control of the process. Sensor are sensitive to changes in the air composition as air is sucked into air chambers in the stationary stations or on top of the odor detecting vehicle. When a change is detected, the sensor triggers an alarm if a defined level is reached or exceeded.

The Ministry of Science and ICT (MSIT) announced in July, 2017 that it would be cooperating with the Ministry of Land, Infrastructure and Transport (MOLIT) to widen the area with IoT technology to apply the SIA solution as a part of Busan-wide smart city regeneration effort.

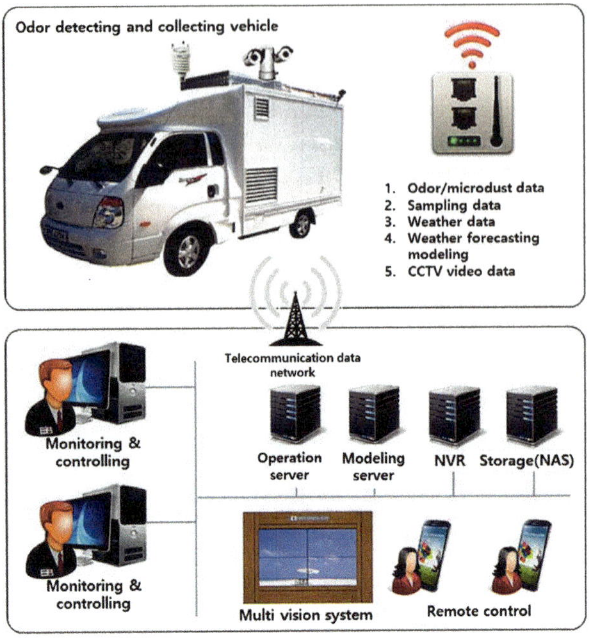

Fig. 3. Odor detecting and collecting vehicle transmitting data to the IOCC. Source: https://www.youngjintech.com/odor-collector

2.2 Other Measures that Helped Reduce Odor Emissions

The IOCC is partly responsible for the spectacular results obtained concerning the decrease of odor complaints in Sasang District. It must be mentioned that the following factors have also played a significant part in the reduction of odors.

- Changes in demographics and land use: The moving out of factories from SIA to other parts of Busan have reduced the number of odor emitting factories. Factory buildings are being demolished and apartment buildings are being constructed in their places.
- Environmental policy to reduce odor: An odor management area was determined by local authorities, supported by legislation for odor prevention.
- Tree planting to absorb odors from the air: 1.5 km long area was heavily planted with 70,000 trees at a cost of 1.8 billion Won.

3 Summary

The increasing attention of residents to odors in the air and the need to provide reliable data have promoted diverse measurement and collection technologies. The case of SIA has shown that these data can be shared and utilized for proactive measures by applying IoT technologies via IOCC. It should be noted that cutting edge technology was applied with more traditional methods, such as legislation and tree planting. These measures have proven to contribute significantly to the control and reduction of odors in the air. The future possibilities are promising as shown by the dramatic results of odor emission detection and control since the operation of IOCC in Sasang District in Busan.

Acknowledgement. The research was supported by the International Science and Business Belt Program through the Ministry of Science and ICT (2017K000451).

References

1. Brattoli, M., de Gennaro, G., de Pinto, V., Loiotile, A.D., Lovascio, S., Penza, M.: Odour detection methods: Olfactormetry and chemical sensors. Sensors **11**, 5290–5322 (2011). https://doi.org/10.3390/s110505290
2. Szulczynski, B., Wasilewski, T., Wojnowski, W., Majchrzak, T., Dymerski, T., Namiesnik, J., Gebicki, J.: Different ways to apply a measurement instrument of e-nose type to evaluate ambient air quality with respect to odour nuisance in a vicinity of municipal processing plants. Sensors **17**(11), 1–19 (2017). https://doi.org/10.3390/s17112671
3. Kim, D.: Sasang gongeupjiyeokeu byeonmowa jeunmang (The transformation and prospects of Sasang Industrial Area). Today's Culture Criticism **60**, 128–142 (2006). (In Korean)
4. Chang, H.Y.: Akchui minwon gangseu neulgo sasangeun juleu (Citizen's complaints of odors increased in Gang-seo district while reduced by half in Sasan district), Gukje Sinmun, vol. 7, 28 November 2016. http://www.kookje.co.kr/news2011/asp/newsbody.asp?code=0300&key=20161129.22007200045. Accessed 31 Dec 2017 (In Korean)
5. Lim, S.Y., Choi, Y.J.: Akchui jadong cheukjeungmang unyeong (Operation of automatic odor measurement network). Annu. Rep. Busan Metrop. City Inst. Health Environ. **26**, 273–282 (2016). (In Korean)
6. Lee, H., Jeonk, S., Choi, W., Lee, S., Lee, M., Oh, K.: A novel assessment of odor sources using instrumental analysis combined with resident monitoring records for an industrial area in Korea. Atmos. Environ. **74**, 277–290 (2013)

The Development of Smart Farm with Environmental Analysis

Panupong Tanomkiat[1], Kobsak Sriprapha[2], Hathaithip Sintuya[1], Nuttiya Tantranont[1], and Worajit Setthapun[1(✉)]

[1] Asian Development College for Community Economy and Technology, Chiang Mai Rajabhat University, Chiang Mai 50180, Thailand
worajit@cmru.ac.th

[2] National Electronics and Computer Technology Center, Pathum Thani 12120, Thailand

Abstract. This work is an on-going research to develop a smart farm system with environmental analysis for terrace vegetable farming with web application. The terrace farm is separated into 3 levels of moisture settings according to the type of vegetables that require low, medium or high moisture content. The environmental sensors monitor the moisture in the soil and send the information to the database of the farm server. The processed information from the sensors will be analyzed and then provide the appropriate environmental factors in plant growth. The smart farm can be monitored and controlled by web application. This smart farm is considered low carbon agriculture due to the power usage from the community microgrid of 25 kW PV system.

Keywords: Database · Farming sensor · Smart farm · Terrace farming

1 Introduction

At present, agricultural farmers faced numerous obstacles such as shortage of labor, shortage of water and low quality agricultural products. In addition, the modern farming in Thailand emits large amount of carbon dioxide which is due to the usage of large farming machineries. The carbon dioxide emission affected the environment. The farmers also have to be responsible for the ever-increasing investment for farming and resulted in reduce income. Therefore, to solve the farmer's issues, there is a need to develop an effective farming practices to achieve high yield and increase the value of agriculture from the limited resources. Technologies can be the solution to the current issues in farming [1]. There are several researches on sensor technologies [2], temperature and irradiation control in green house [3], development of environmental sensors and data transfer through wireless protocol [4, 5]. These technologies could be applied to the smart farm system to control the farming condition to provide high yield plant production.

This research main objective is to develop the smart farm system in the small community level. The analysis of farm environmental conditions will determine the factors affecting plant grown. Database of farm environment factors can be used to analyze and control the farm through the web application. The goal is to increase the

S. O. Hwang et al. (Eds.): ICGHIT 2018, LNEE 502, pp. 210–214, 2019.
https://doi.org/10.1007/978-981-13-0311-1_35

quality and quantity of products and reduce the farm production cost. In this work, the smart farm is integrated with the power generation system from 25 kW PV microgrid system.

2 Smart Farm Design

The smart farm prototype was set up at the Smart Community Farm of the Asian Development College for Community Economy and Technology (adiCET), Chiang Mai Rajabhat University. The terrace farming area was 200 m^3.

2.1 Terrace Farming Area

In this work, terrace farming was set up with the moisture monitoring system to control 3 level of terrace. The moisture levels of the soil were set as 3 levels: low, medium and high.

2.2 Smart Farm System

The Smart Farm System comprised of 4 parts which are Vegetable Garden, Weather Station, Server & Database, and Monitoring & Control through Web Application. Figure 1 showed the Block diagram of the 4 parts.

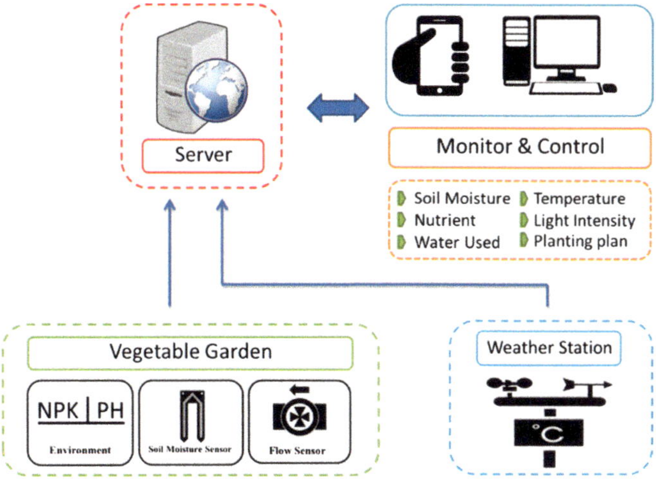

Fig. 1. Block diagram overview of Smart Farm

Vegetable Garden

Vegetable garden will install the soil moisture sensors, water flow sensors, NPK sensors and pH sensors. The data from these sensors will provide the real-time environment information of smart farm. The data will be then transfer to database in the farm's server.

Weather Station
Weather station comprise of light intensity sensor, relative humidity sensor, temperature sensor and wind speed sensor. The information will provide the trend of environmental changes at the smart farm and also the calculation of the reference crop evapotranspiration from the environmental data [6].

Server
Server stores the database of the farm data from the vegetable garden and weather station for analysis.

Monitoring and Control
Monitoring and control focuses on the status display and control of the smart farm through web application. This function can control water distribution in the farm.

2.3 Data Transfer

The MQTT protocol is a way to send data to the MQTT Server, which is the intermediary for receiving the information from the sensors in the farm. As shown in Fig. 2, the moisture, environmental and flow sensors will measure and send data to the server. The server is responsible for processing the data received from sensors on the farm and working with the MySQL database. The processed data will be displayed and the farm conditions will be controlled through web application.

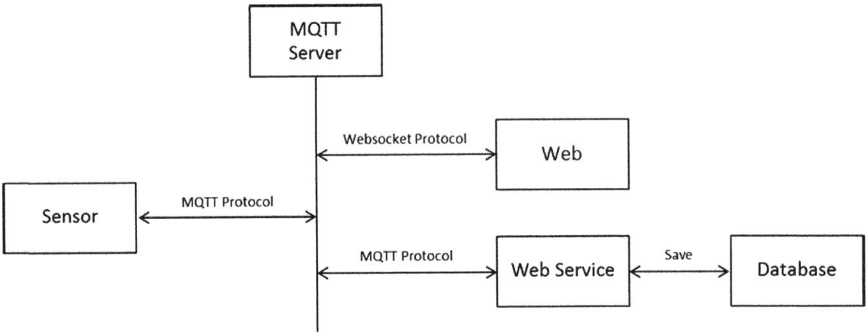

Fig. 2. Transmission of sensor data within the farm

3 System Database

The smart farm system database is categorized into 3 parts which are (1) Suitable factors for plants database which is the reference database for Soil moisture, Nutrient, pH, Light intensity and Temperature that are suitable for each plant; (2) Environmental in agricultural database which will stored the real data from the terrace smart farm with the data of Soil moisture, Nutrient, pH, Water usage; and (3) Weather data from the weather station which includes data, Light intensity, Relative humidity, Temperature and Wind Speed.

For the analysis of the environment for plants, the real data from the sensors is compared to the suitable factor for the plant database. The farm environmental data will be compared with the weather database. The comparison will be processed and the results can be the recommendation to control the farm conditions to achieve suitable condition for each type of plant (Fig. 3).

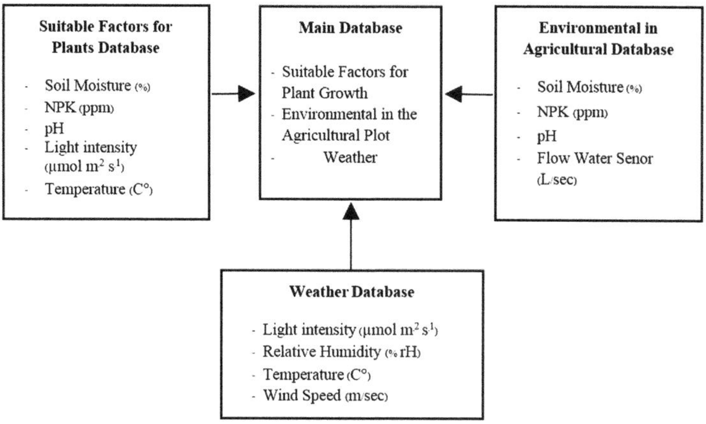

Fig. 3. Database format

4 Results

This is an on-going research. At present, the farm moisture sensors and water distribution system were installed. The power is from the 25 kW PV system near the smart farm. The farm was arranged into 3 terrace levels for low, medium and high moisture setting. The Moisture Control System Performance was tested by measuring the moisture of 10 points in the farm with 10 repetitions for each point. The average moisture measurement results from 3 plots according to the 60%, 75% and 90% moisture settings were found at 62.26%, 76.84% and 92.19%, respectively. The error of the soil moisture measurement system was at ±2.19%. These data will be transferred to the agricultural database via MQTT protocol and will be used for the smart farm processing to develop the suitable conditions for each type of plants.

5 Conclusion

The smart farm with environmental analysis focused on the development of measurement system and the suitability of the environment analysis in farm. Measurements of the farming condition will be collected to establish a database of environmental factors that affect the growth of plants. This data will be the basis for comparative analysis of environmental factors suitable for plant growth.

Acknowledgements. Asian Development College for Community Economy and Technology for the experimental site location and Thailand Graduate Institute of Science and Technology for the funding.

References

1. Phromchaipattana, T., et al.: What is stalling the progress of Thailand agriculture. Adv. Bus. Mag. **27**(325), 2–3 (2016)
2. Darshna, S., Sangavi, T., Mohan, S., Soundharya, A., Desikan, S.: Smart irrigation system. J. Electron. Commun. Eng. **10**(3), 32–36 (2014)
3. Kanjilal, D., Singh, D., Reddy, R., Mathew, J.: Smart farm extending automation to the farm level. Int. J. Sci. Technol. Res. **3**(7), 109–113 (2014)
4. Roham, V., Patil, A., Pawar, G., Rupnar, P.: Smart farm using wireless sensor network. Int. J. Comput. Appl. **7**, 8–11 (2015)
5. Zhu, Y., Song, J., Dong, F.: Applications of wireless sensor network in the agriculture environment monitoring. Procedia Eng. **16**, 608–614 (2011)
6. Irrigation Water Management Division: Crop Water Requirement Reference Crop Evapotranspiration & Crop Coefficient Handbook. Royal Irrigation Department, pp. 22–58 (2011)

Multi-modal Sensor Calibration Method for Intelligent Unmanned Outdoor Security Robot

Taeyoung Uhm(✉), Gi-Deok Bae, JongDeuk Lee, and Young-Ho Choi

Field Robotics R&D Division, Korean Institute of Robot and Convergence, Pohang, Republic of Korea
{uty, bgd9047, artofgene, rockboy}@kiro.re.kr

Abstract. Robots designed for outdoor use in an outdoor environment should be able to autonomously travel using multi-modal sensors. For autonomous navigation, it is a need for a calibration method that employs a multi-modal sensors for integrates all sensor coordinate systems. In this paper, we propose a corporate calibration method using a multi-modal sensor module. For this purpose, we use a grid pattern can be extracted from various sensor data and the coordinate system for each sensors is integrated based on the correspondence of extracted corner points using pattern information. The proposed framework has been tested extensively in actual outdoor scenes with our self-developed multi-modal sensor module. The experimental results validate its usefulness and efficiency.

Keywords: Multi-modal calibration · Multi-modal sensor · Outdoor robot

1 Introduction

Recently, interests in the intelligent robotics are increasing. Especially robots designed for surveillance in the outdoor environment are expected to play a major role in the industry. Autonomous driving and mission execution are essential elements for these robots, and many researchers are conducting studies on them. For regulating navigation and mission, environment recognition and localization should be possible by using multi-modal/multi-type sensors. In addition, since most of the sensors employed for the recognition have respective coordinate systems, it is necessary to integrate them into one coordinate system. In this paper, we propose an apparatus and a calibration method for integrating these multi-modal and polymorphous sensors into a single coordinate system.

In general, the calibration method finds some corner points of a pattern in a plurality of camera images and performs correction for each corresponding points [1]. This method should mainly use mono camera or color camera images with a pattern. On the other hand, a method of performing correction between different types of camera images has also been proposed [2]. However, this study is possible because it is able to find the grid pattern in the images for each camera sensor, and it is impossible in the

S. O. Hwang et al. (Eds.): ICGHIT 2018, LNEE 502, pp. 215–220, 2019.
https://doi.org/10.1007/978-981-13-0311-1_36

sensor which cannot find the grid such as the thermal image. The calibration method for thermal images is proposed [3], but it is extremely limited.

In contrast to the image-based method, a calibration method for sensors (e.g. 3D LiDAR) have also been proposed [4]. However, this method is also calibration application which is not suitable for other types of sensors. Therefore, there is a need for a calibration method that can perform correction of RGB image, night vision image, thermal image, and 3D LiDAR sensor data, which are multi-modal sensors necessary for autonomous driving and mission execution. By this method, the ability to understand the robot's surroundings is possible, and the robot performance heavily depends on the reliability of its environment perception technologies including self-localization. Consequently, we develop a device that can find the corner point in multi-modal sensors (RGB image, thermal image and 3D LiDAR data), and propose a method to integrate it into one coordinate system using it.

2 Multi-modal Sensor Calibration Method

2.1 Multi-modal Sensor Module

For the outdoor unmanned security robot, a multi-modal sensor module is de-signed to share the field of view (FOV) for each sensor (Two RGB cameras, a Night vision camera, a Thermal camera and 3D LiDAR etc.) that can be acquired and integrated into one module considering portability in outdoor environments, as shown in Fig. 1.

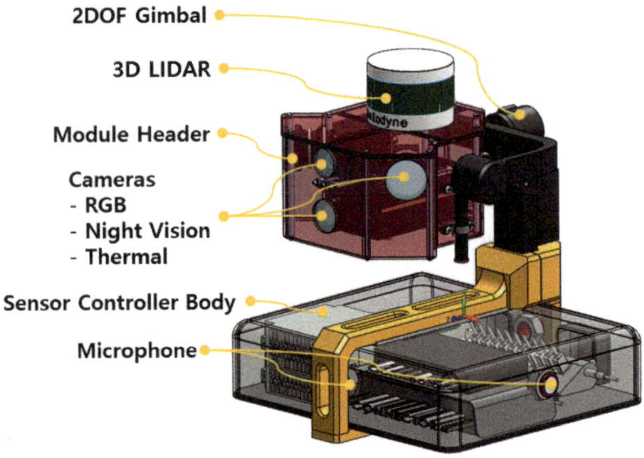

Fig. 1. The multi-modal sensor module for outdoor robots.

In order to integrate the coordinate system of each sensor data acquired from the multi-modal sensor module, a device capable of sharing the corner information of pattern in each sensor is required. For this purpose, we build a pattern board which can find the correspondences of corner points in various sensors. The board has built-in white and black grid patterns for RGB images and lines visible in thermal imaging to

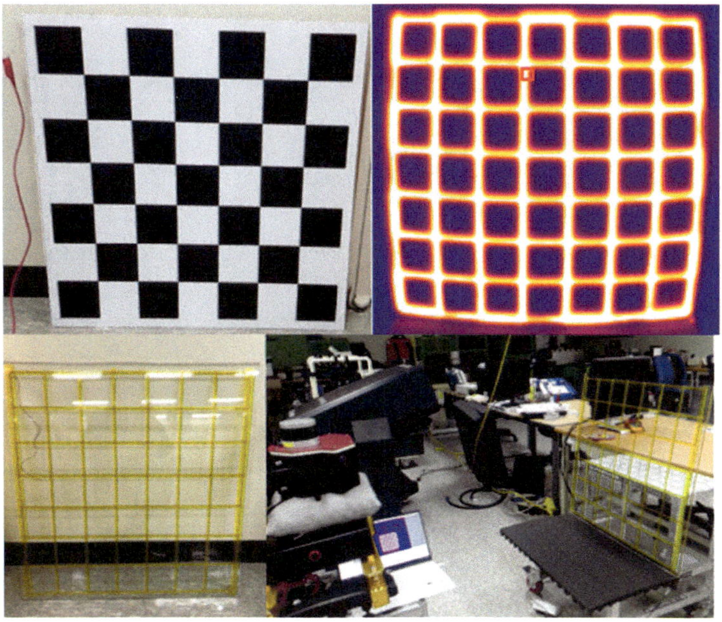

Fig. 2. A pattern board for multi-modal sensor module.

the edges of the grid pattern. In addition, we use the edge points of the pattern board for 3D LiDAR calibration. Figures 3 and 4 show images obtained by the multi-modal sensor module. Then, each corner point has extracted from the images. These can be used to perform calibration with a single coordinate system (Fig. 2).

2.2 Multi-modal Sensor Calibration

The single coordinate system has a difficulty by organization required to correct each sensor. We employed a systematic and stepwise approach using multi-modal sensor module. By the camera sensor coordinate, a reference image can be transformed to another image by rotation matrix R and translation matrix t. Therefore, a point xw from world coordinate is represented by

$$x_c = R_c x_w + t_c \tag{1}$$

where xc is an camera image point. For the combination of two camera sensor coordinate, we use a well-studied problem; Zhang's method [5] to calibrate by checkboard corners. Next, a homography is computed for each images using the corner positions in world coordinate. Then, each homography H extract camera parameters that are solved by linear equations by

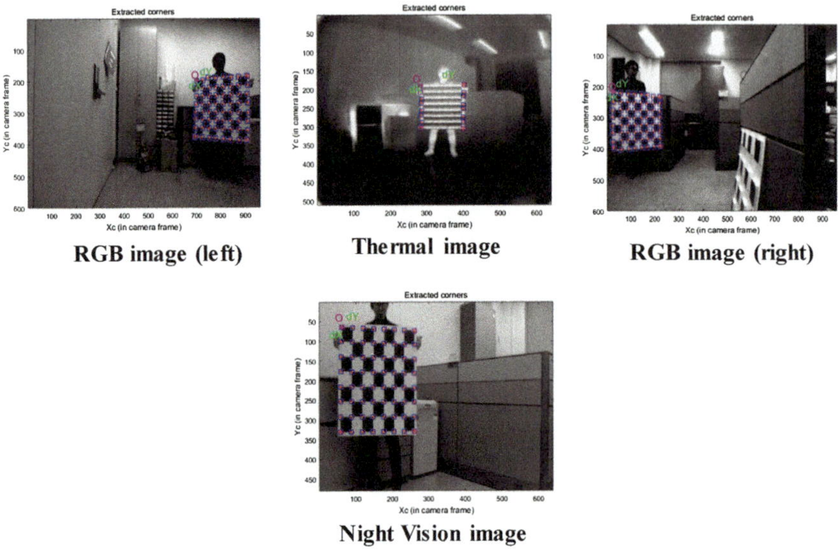

Fig. 3. Multi-modal sensor images with corner detection.

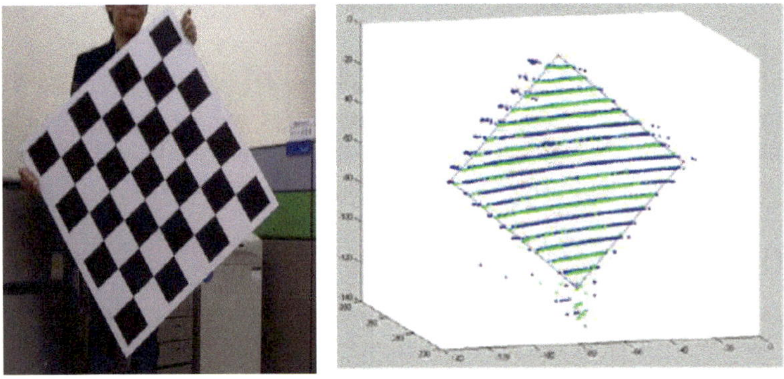

Fig. 4. Corner points by 3D LiDAR.

$$H = A[R_c\ t_c] \tag{2}$$

where A is camera intrinsic matrix. The relative camera pose can be obtained directly because the pose has the same reference frame is known. For the proposed multi-modal sensor system, night vision camera sensor is central. Therefore, we extract each homography for the pair that consists of night vision and i-th camera as following:

$$H_{mi} = H_n^{-1} H_i. \tag{3}$$

The homography can be estimated each initial pose from the night vision camera in a single coordinate system. Finally, we calculate a homography between night vision and 3D LiDAR sensor using four corners.

3 Experimental Results

Figure 5 shows the calibration results for the pair of cameras. After correcting three relative coordinate systems, there are integrated into a single coordinate system centered on night vision camera. Finally, we constructed the coordinate system which is incorporates a 3D LiDAR coordinate system.

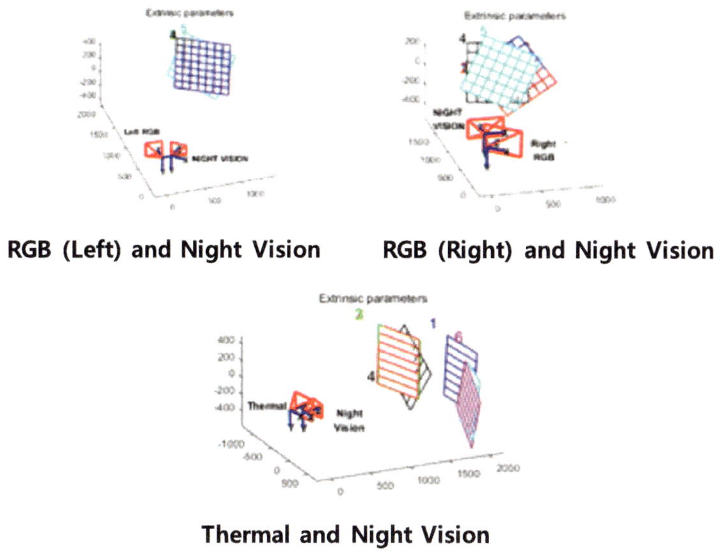

Fig. 5. Calibration results of each cameras.

4 Experimental Results

This paper presents a calibration method for multi-modal sensor module which designed for intelligent unmanned outdoor security robot. The calibration method takes advantage of multi-modal/multi-type data for the sensor module. This research expects the method can be applied to a typical robotics application for autonomous driving and mission execution.

We are currently implementation various robot applications. In the future, we are going to analyze and perform an accuracy improvement on this method.

Acknowledgement. This work was supported by the Korean Evaluation Institute of Industrial Technology and conducted by the Ministry of Industry and Commerce in 2017 (Industrial Core Technology Development Project, Project Number 10080489).

References

1. Clarke, T.A., Fryer, J.G.: The development of camera calibration methods and models. Photogram. Rec. **16**(91), 51–66 (1998)
2. mm-Calibrator—OpenCV-Based Multi-Modal Camera Calibration System—Google Project Hosting. http://code.google.com/p/mm-calibrator
3. Prakash, S., Lee, P.Y., Caelli, T., Raupach, T.: Robust thermal camera calibration and 3D mapping of object surface temperatures. In: Miles, J.J., Peacock, G.R., Knettel, K.M. (eds.) Thermosense XXVIII, SPIE, Bellingham, WA, p. 62050J (2006)
4. Zhang, Q., Pless, R.: Extrinsic calibration of a camera and laser range finder (improves camera calibration). In: Proceedings of the IEEE/RSJ International Conference on Intelligent Robots and Systems, vol. 3, pp. 2301–2306 (2004)
5. Zhang, Z.: A flexible new technique for camera calibration. IEEE Trans. Pattern Anal. Mach. Intell. **22**(11), 1330–1334 (2000)

User Controls in Video Regeneration System

Young-bong Kim, Yoseph Yoon, Keon-kuk Park, Oh-Seok Kwon, and Jongnam Kim(✉)

IT Convergence and Application Engineering, Pukyong National University, Yongso-ro, 45, Nam-gu, Busan, Republic of Korea
{ybkim, jongnam}@pknu.ac.kr

Abstract. Recently, a variety of multimedia materials have been rapidly generated to generate new data using various blending techniques. The commercial video blending technology used in movies, broadcasts, and commercials requires expertise and is complex for end users. In this paper, we propose a recycling system which recycles images that can obtain similar image synthesis results with less effort than conventional commercial programs. The proposed system consists of three stages: video classifier, image search extractor, and video blender. The composite image generator is implemented with the emphasis on the synthesis of moving objects and semi-fixed objects, such as objects, faces and background compositing. The system uses Poisson Image Matting techniques to handle natural border changes in images and images, feature techniques such as Haar to automatically track face location, size and angle, And a watershed partitioning scheme. The proposed system is easy to use by untrained people because it can produce high quality blended images with simple manipulation of video and image compositing.

Keywords: Video blending · Poisson image matting · Haar-line feature Watershed segmentation

1 Introduction

Recently, as media such as SNS, smart phone, tablet, and Internet TV are developed, it is not aimed at people who are providing content, but it is moving towards providing personal goal [1]. In addition, with the development of real-time streaming protocols, more and more producers are increasingly trying to provide content for individuals or a small number of people [2]. In addition to creating and delivering content that meets the public's needs, you need to move to a custom content creation system that can reflect your ideas. However, most commercial software used in film, broadcast, and commercial production is well suited to those who have acquired specialized skills. Although commercial programs can easily display subtitles or images, blending between objects, faces, and backgrounds requires a lot of manipulation to naturally synthesize. Common commercial programs have the following effects.

Motion tracking technology to extract motion [3], matte choker effect [3] to soften blending boundaries, and curve filter to obtain similar colors [3]. To get natural shadows there is Drop Shadow effect [3]. In this paper, we propose a method to generate a similar effect through one or two of the above effects. And we have designed

S. O. Hwang et al. (Eds.): ICGHIT 2018, LNEE 502, pp. 221–226, 2019.
https://doi.org/10.1007/978-981-13-0311-1_37

and implemented a simpler video playback system in a suggested way so that end users can more easily create custom video content.

1.1 Video Regeneration System Overview

In this paper, we propose a video reconstruction system based on original source video created by content providers and image extracted from candidate image obtained through tag-based searching. And a new video is finally generated through a combining process of correcting them. The proposed system consists of three steps as shown in Fig. 1. First, a video classifier which distinguishes the frames to will be blended in the original video source. Second, an image search extractor that search a video to be blend and extracts a necessary image from the video. Finally, video blender that regenerates a new video blended with original frames and the image through frame comparison and correction. In addition, the video blender provides three main functions: object blending for inserting a new object into the original video frames, face blending for replacing original face from original frames with the new face from the extracted image, and background blending for replacing the background of the original video with extracted image.

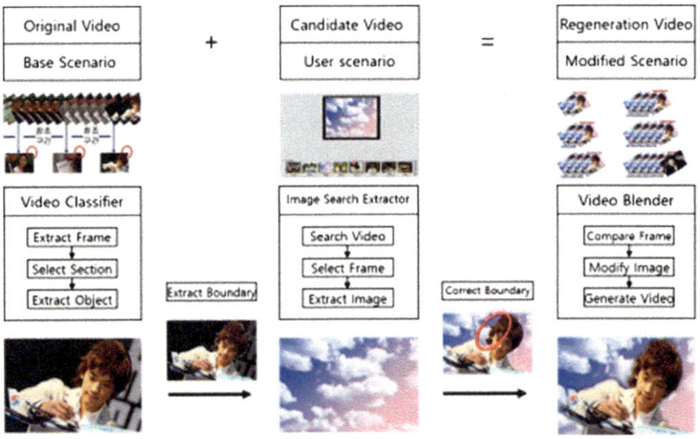

Fig. 1. System overview

2 Design and Its Implementation

In this chapter, we will explain the 3 stages of video regeneration system step by step. In addition, it also refers to the techniques to blend objects, faces, and backgrounds, respectively.

2.1 Video Classifier

In *Video Classifier*, first, it splits reference sections have frames composed of similar images from an original source video. Second, based on scene change, it is indexed as the candidate frames from scene change section. Third, it selects the changing section include the background range or objects of interest, and then, Lastly, extracts frames will be changed from the reference section (Fig. 2).

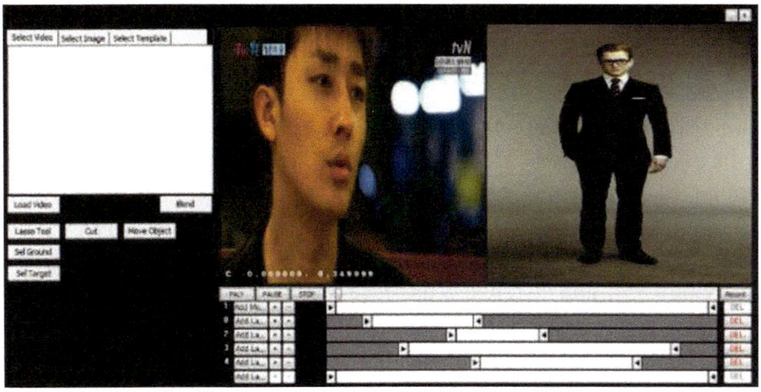

Fig. 2. Video classifier

2.2 Image Search Extractor

In *Image Search Extractor*, the user input text and tags of various videos stored in the candidate video storage are compared with each other, and candidate videos are extracted first. Then, the user selects a desired frame from the video, and extracts a target image for blending the original source video (Fig. 3).

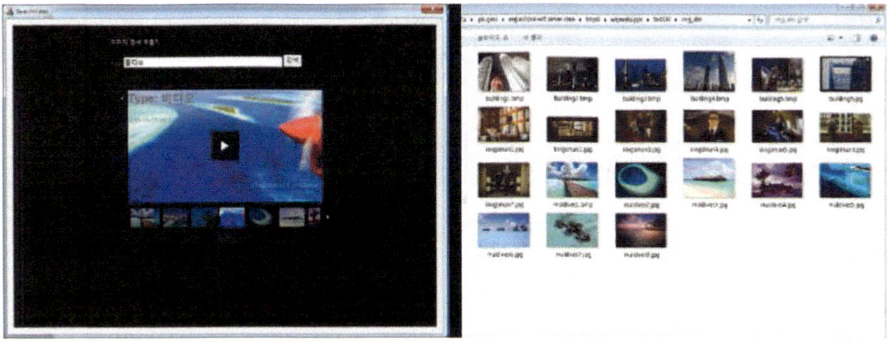

Fig. 3. Image search and extractor

2.3 Video Blender - Object Blending

Video blender includes three effects; object blending, face blending and background blending. Here, an *Object* refers to an object other than a face, a background, and the like among the objects constituting the video. The result should show the desired object in the video. The manipulation is as follows. First, a video and an image are extracted through the above steps. Second, the image is placed on the blending position of the video and rescaled by user interaction. After that, the *Poisson Image Matting* technique was applied to blend objects automatically, and the boundary was processed smoothly and naturally [4].

2.4 Video Blender - Face Blending

Face blending should automatically extract a face from the original image and replace it to another face image extracted from above steps. The manipulation is as follows. First, a video and an image are extracted through the above steps. Then, the extracted image is mapped to the 3D face model. After that, the system automatically employs *Haar-like Feature* algorithm to automatically detect the position, angle and size of the face in the original video [5], and it replace the face with the 3D face model and blending them.

2.5 Video Blender - Background Blending

Background blending separates background and foreground from the original image, and then remove the background, finally blend the extracted background to the original video instead of the existing background. The manipulation is as follows. First, draw two lines to separate the background and the foreground. After that, it automatically employs *Watershed Segmentation* technique to segment foreground and background to each region [6]. The background region is removed, and then original background image is replaced with another extracted background area (Fig. 4).

Fig. 4. Background blending

3 Experimental Results

To verify the quality of the proposed system, we compared the results with *Adobe After Effects*, one of the commercial programs. Figure 5(a) shows object blending result of our system, and Fig. 5(b) represents object blending of the after effect. Proposed system can obtain similar result by *After Effect* employing *Matte Choker Effect* and *Curve Filtering*. However, it cannot give shadow effects. Figure 5(c) and (d) are the background blending of our system and *After Effect*, respectively. Figure 5(e) is our face blending, and Fig. 5(f) shows face blending by *After Effects*. Figure 5(e)-1, (f)-1 are extracted face images, and (e)-2 and (f)-2 are a frame of original video. (e)-3, (e)-4, (f)-3 and (f)-4 represent result of image that replacing original face with extracted face image. As shown in Fig. 5, (f) face blending extracts the movement of the face of the original video through Motion Tracking and employ the new face image the movement. However, Fig. 5(e) our face blender can extract movements automatically by *Haar-like Feature* algorithm, but it cannot be used when there are more than two faces. As shown in Fig. 5, the results of the commercial system are largely different in quality depending on the user's skill. On the other hand, this system has confirmed that the user's ability has little effect on the quality of the result, but the similar effect to the commercial system with little manipulation.

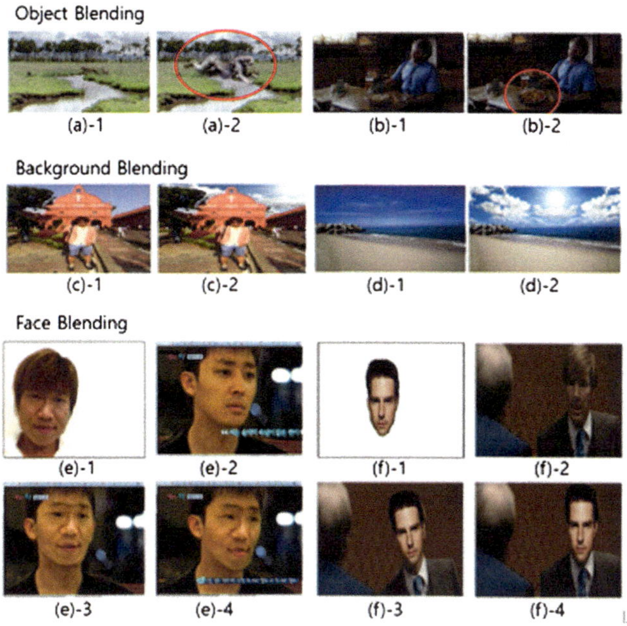

Fig. 5. (a) Object blending of proposed system, (b) object blending of the after effect, (c) background blending of proposed system, (d) background blending of the after effect, (e) face blending of our system, (f) face blending of the after effect

4 Concluding Remarks

The number of customized video content producers has increased. However, current commercial systems are inconvenient to use for unskilled people. Because there are many manipulations to obtain natural results in video blending. To satisfy the needs of the individual video creating, we propose a video regeneration system that can produce similar effects to commercial system while minimizing manipulation. The blending boundary automatically have softened by the Poisson image matting technique. In the case of face blending, the movement of the face in an original video automatically have tracked through the Haar-like feature algorithm. In case of background blending, we could distinguish foreground and background area by simple operation through Watershed Segmentation. We think that our system can be helpful for individual content producers.

Acknowledgment. The work was supported from PKNU research year project and URP of PKNU.

References

1. Lee, D.Y., Lee, S.Y.: Media experience in live streaming video service: comparative study on parasocial interaction and social presence among live sports streaming video service users. Korean J. Journalism Commun. Stud., 148–177 (2014)
2. Choi, M.Y.: Personal broadcasting service trends and prospects. J. Korean Inst. Commun. Sci., 66–70 (2016)
3. Adobe Support. https://helpx.adobe.com/
4. Sun, J., Jin, J., Tang, C.L., Shum, H.Y.: Poisson matting. ACM Trans. Graph. (ToG) **23**(3), 315–321 (2004)
5. Lienhart, R., Maydt, J.: An extended set of Haar-like features for rapid object detection. In: Proceedings of the International Conference on IEEE Image Processing (2002)
6. Couprie, C., Grady, L., Najman, L., Talbot, H.: Power watersheds: a new image segmentation framework extending graph cuts, random walker and optimal spanning forest. In: 12th IEEE International Conference on Computer Vision (2009)

An Evaluation to Firmware Code Materials for Embedded IoT Device

KyuTae Lee[1], HyunChang Lee[1(✉)], Sung Yeol Kwon[2], Sangyep Nam[3], and DoHyeun Kim[4]

[1] Division of Information and Communication Engineering, Kongju National University, Kongju, South Korea
{ktlee,hclee}@kongju.ac.kr

[2] Department of Electrical Engineering, Pukyung National University, Busan, South Korea
sungyeol@pknu.ac.kr

[3] Department of Computer IC Engineering, KookJe University, Pyeongtaek, South Korea
synam59@gmail.com

[4] Department of Computer Engineering, Jeju National University, Jeju City, South Korea
kimdh@jejunu.ac.kr

Abstract. A portable IoT device has a hardware system configured with a proper processor and software program code. A source code evaluation is a method of resolving to an illegal modification between the original code and the suspected program code on copyright software. The evaluated result is extracted from the comparison of the each assigned item to seeking objective facts. On the verification process, program experts compare both codes to the requested items from the copyright association. Therefore a detailed comparison item has a leverage as to it has a sensitive value in decision. In this case study, we analyzed and verified the object firmware code with the detailed item selection and suggest an additive comparison item in embedded IoT device.

Keywords: Source code · Copyright · Ownership · Comparison

1 Introduction

The digital contents based on IT software is gradually expanded and increased with its usability, producing and distribution with easy to every area. But there are difficult work to protect an illegal copy and transformation and many issues in copyright protection to keep the original ownership of contents developer against attacks. Also it can be decreased the creative activity of contents designer and even issued to criminal case as to obstructing new contents product. So copyright protection is becoming an important issue to an ownership protection and is growing up for keeping creative contents of original producer from illegal activities. In the domestic organization, the copyright association, is now taking a charge into copyright protect and contents verification by comparison to suspected materials, when there is conflict issues in

S. O. Hwang et al. (Eds.): ICGHIT 2018, LNEE 502, pp. 227–231, 2019.
https://doi.org/10.1007/978-981-13-0311-1_38

ownership [6]. The processing of comparison is a work for extracting a meaningful numerical likelihood throughout matching process between original and suspect contents. In this paper, we verified and suggested the way to adapt a reasonable comparison process and extract a likelihood of both contents to actual Object materials in embedded IoT device as a case study.

2 Device Basics

General microprocessor system device has an architecture included CPU, FlashROM, SRAM, and peripheral interfaces in Fig. 1. For embedded system having operation system, the processor has a Flash memory and special interface block on a mobile service [2].

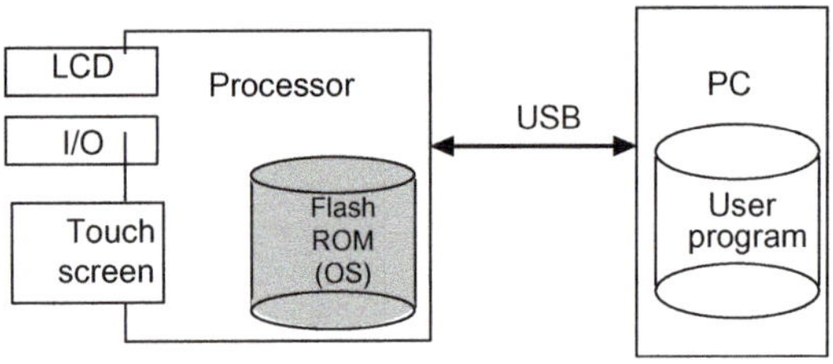

Fig. 1. Configuration of general embedded system

What we are interesting is a source code which has a system program. The system program has an application code with an operation system code, such as Linux or Windows [1]. As a case study, an object material is a touch screen device which sense a touch signal as a key technology, such as IR (Infra Red) sensing for large scale screen and resistive and capacitive sensing for small scale like PDA or smart phone. The IR sensing is useful for large screen cause it has no pad on the LCD display and needs more power and system size. The device is figured to ARM7 processor and installed USB interface getting a location value touched and saving and sending the data. These interface program code is programmed with C language and saved in FlashROM as a firmware [3, 4]. An application code and interface driver code are configured to work in PC environment as Fig. 2.

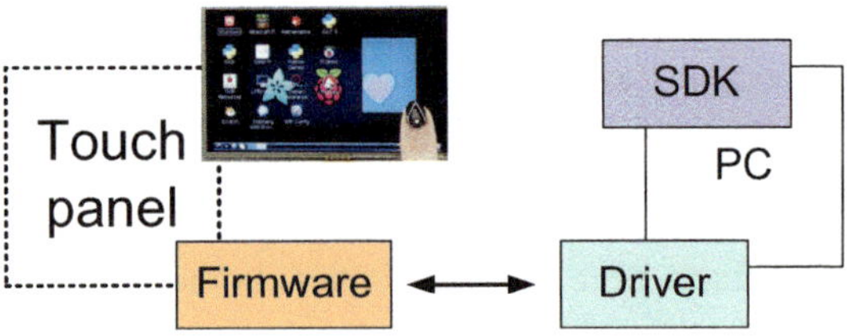

Fig. 2. Embedded processor architecture

For the evaluation, application program is compared in a working sequence practically. And codes of Firmware code and driver, user interface code are compared to general evaluation process.

3 Code Evaluation

The procedure of searching similarity from object materials is shown on Fig. 3. At first, a person who has disadvantages on his properties accuses a person who did an illegal copy to investigatory agency. The agency begins to investigate the materials submitted from both sides. Then after ask an expert opinion such as copyright association. In this step, if it can be pointed out a specific part to investigate, it is possible to assign specific search items to copyright association with agreement. The evaluation expert takes an investigation for the similarity to the specific item or total items as requested.

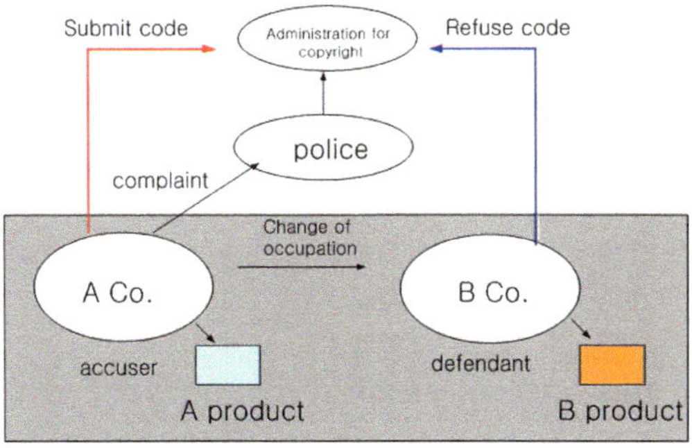

Fig. 3. Compare processing with both materials

For the evaluation task, assigning a subtitle item is important for reducing an objective result and applying a weight factor to each item. A designation for evaluation experts is assigned by the copyright association for analyzing the subjected materials. Also searching for open source to specific items is performed to find out an originality of the property owner on the materials [6].

On doing analysis, the decision depends on the evaluation expert. He deduced the original evidence and tried to find out which kinds of knowledge used: such as common or specific technologies.

Generally, illegal program is constructed to a way to modification from original code source cause that is time consuming module and it has a key technology code block which is hard to make. Also they add a new program source code block to original program which has a new idea of property owner. Sometimes besides key function block of original code, there are trials to hide illegal acts by modifying easy interface blocks such as GUI or display colors etc [5]. Therefore a suspect code based evaluation method is not correct and the likelihood value does not have critical meaning to make decision. So basically source code based evaluation is accepted and exceptionally suspect code based method is adapted when original code is included to the suspected code material with too many part. Finally there is no principle which method is better, that is mainly the right to evaluation expert decision under conflict material inspection.

After evaluation, similarity value of firmware code shows 10.9% even with a same material comparison trial as below Table 1. And it showed a 54.8% average value which has driver and application code likelihood. Here we added a logical compare process to find an original technology with practical operation sequence on both materials.

Table 1. Result of firmware code

	Firmware	Single firm lines	Multi firm lines	Revised lines	Revision ratio (%)
1	interrupt_timer.c	292	292	0	0
2	dc_enumerate.c	946	946	51	5.40%
3	dc_enumerate.h	45	45	0	0
4	dbgu.c	174	174	0	0
5	dbgu.h	30	30	0	0
6	ext_irq.c	64	64	0	0
7	Flash.c	261	261	0	0
8	Flash.h	69	69	0	0
9	interrupt_timer.c	1349	1726	337	25.00%
10	interrupt_Usart.c	296	326	10	3.40%
11	main.c	1175	1337	115	9.80%
	Average	4701	5270	513	10.90%

We found a key technology on the interface between UBS communication on the data packet assign structure. As a result, the suspected material has no evidence of an illegal copy and the device is constructed with legal idea.

4 Conclusions

IoT device has a characteristic of interfacing with hardware and software functions. So the comparison of the object materials has to include the ownership of interfacing functions with hardware added with program code. In this study, we deal with a source code evaluation and partial similarity under practical sequence. On searching originality facts of the ownership on source code, the evaluation expert be a person who has a career to object development for few years so as to taking original fact. This paper shows a case study to approaching to exclude a similarity on source code and operation sequence method in embedded IoT device. It is expected for this result to refer to keeping copyright ownership for program developer.

Acknowledgments. This research was supported by Institute for Information & communications Technology Promotion (IITP) grant funded by the Korea government (MSIT) (No.2012-0-00265, Development of high performance IoT device and Open Platform with Intelligent Software), and this research was supported by the MSIT (Ministry of Science and ICT), Korea, under the ITRC (Information Technology Research Center) support program (IITP-2017-2016-0-00313) supervised by the IITP (Institute for Information & communications Technology Promotion) Any correspondence related to this paper should be addressed to DoHyeun Kim.

References

1. Gupta, R.K.: Introduction to embedded system. In: ICS 212, 2002 Winter Workshop (2002)
2. Lee, K.-T., Lee, H.-C., Ki, J.-G.: Establishment of the subtitle on materials for evaluating intellectual ownership. Int. J. Sig. Process. Image Process. Pattern Recogn. **10**(9), 79–88 (2017)
3. Getting started with HBE-EMPOS II, Hanback electronics (2004)
4. Intel PXA255 Processor Developer's Manual
5. Yiu, J.: ARM Cortex-M3 guide, ITC (2011)
6. Copyright Association Homepage. http://www.copyright.or.kr. Last accessed 11 Jan 2017

Faces Recognition Using HAARCASCADE, LBPH, HOG and Linear SVM Object Detector

Jae Jeong Hwang[1], Young Min Kim[2], and Kang Hyeon Rhee[2(✉)]

[1] Kunsan University, Gunsan, Korea
hwang@kunsan.ac.kr
[2] Chosun University, Gwangju, Korea
kimym0702@chosun.kr, khree@chosun.ac.kr

Abstract. To solve the illumination problem of the conventional face recognition system using Haarcascade algorithm, LBPH is merged into the system with the HOG linear SVM object detector, in this paper.

Keywords: LBPH · Haarcascade · HOG +SVM · Multiple face detection

1 Introduction

Various biological recognition system has been developed so far. Many of them such as face recognition, finger print recognition, and iris recognition are already on the market. Those human feature recognition can be implemented on any OS, e.g., Window, Linux, Android, and IOS. In this paper, we developed and enhanced the HOG (Histogram of Oriented Gradients), Haarcascade, and LBPH algorithms using Python language in Windows 7 environment.

2 LBPH Based Haarcascade Scheme

The conventional eigenfaces face recognition and Fisherfaces face recognition are confront with problems for detecting side view face images. Moreover, Fisherfaces face recognition is not perfect for the illumination change in various real condition [1, 6–9].

Figure 1 shows that the Haarcascade scheme is based on Haar wavelet, integral image analysis, Adaboost and cascaded classifier.

Figure 2 shows face detection results for slightly slanted face images. Two rectangles on the images mean incorrect performance of Haarcascade system in case of non-front view images. This problem has also been reported in the literature [2, 3, 8, 9].

Therefore, we developed LBPH based face detection scheme as shown in Fig. 3, consisting of LBP histograms array and distance measure blocks. The scheme works well to detect correct faces as shown in Fig. 4, showing a rectangular region of face even in slanted images.

S. O. Hwang et al. (Eds.): ICGHIT 2018, LNEE 502, pp. 232–236, 2019.
https://doi.org/10.1007/978-981-13-0311-1_39

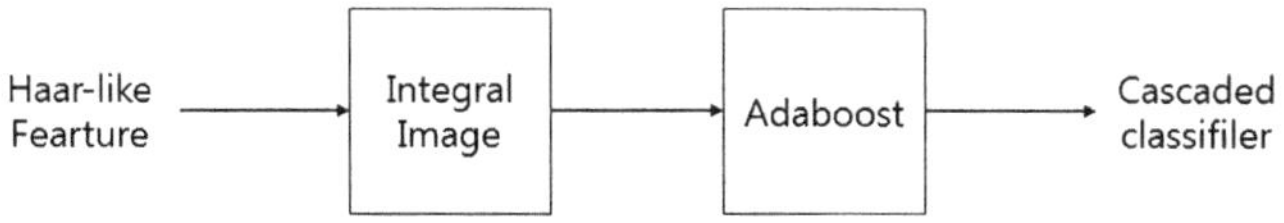

Fig. 1. Face detection block diagram using Haarcascade method.

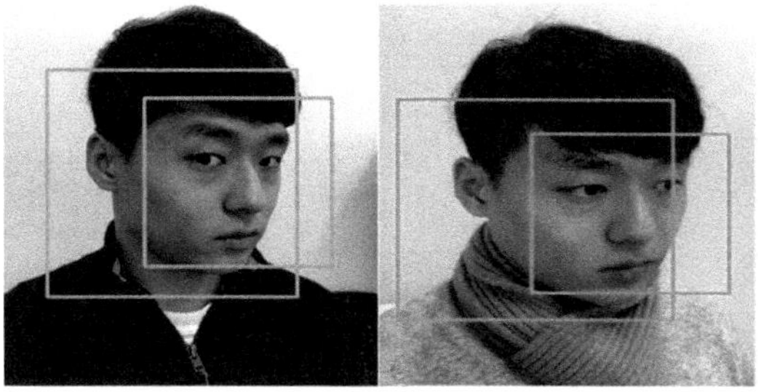

Fig. 2. Face detection example using Haarcascade scheme.

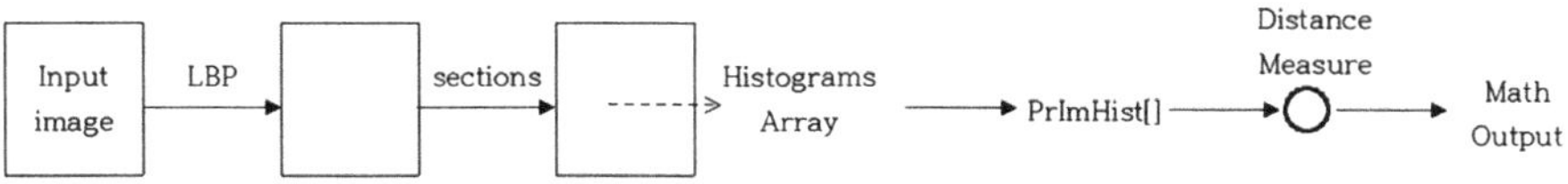

Fig. 3. Block diagram of LBPH based face detection scheme.

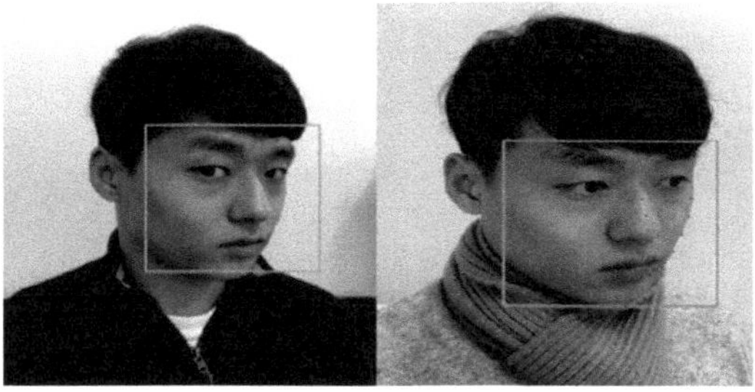

Fig. 4. Face detection results of LBPH based Haarcascade scheme

3 Prepare Your Paper Before Styling

In this work, more enhanced algorithms are proposed as shown in Fig. 5, consisting of LBPH, HOG, and CVM detector. HOG is implemented using OpenCV dlib library. Face feature detector is pre-trained to derive 68 coordinates so that can be mapped on the face boundary as shown in Fig. 6, which is an example for Lena image.

Figure 7 shows two rectangles and mapping points overlaid on slanted images using the conventional Haarcascade method that the detection failure should be solved.

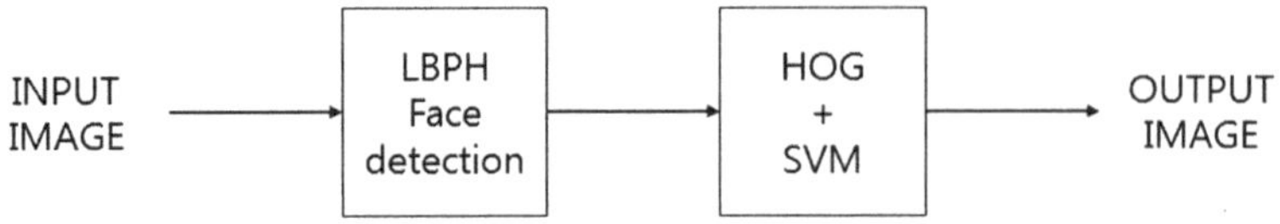

Fig. 5. Block diagram of LBPH, HOG, and SVM block diagram of LBPH, HOG, and SVM.

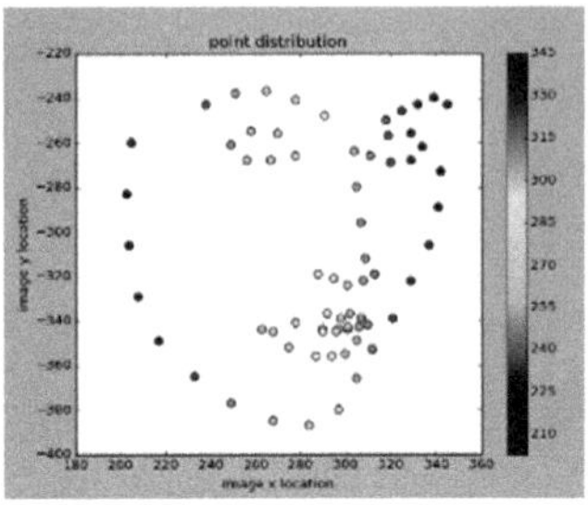

Fig. 6. Mapping points using LBPH based scheme.

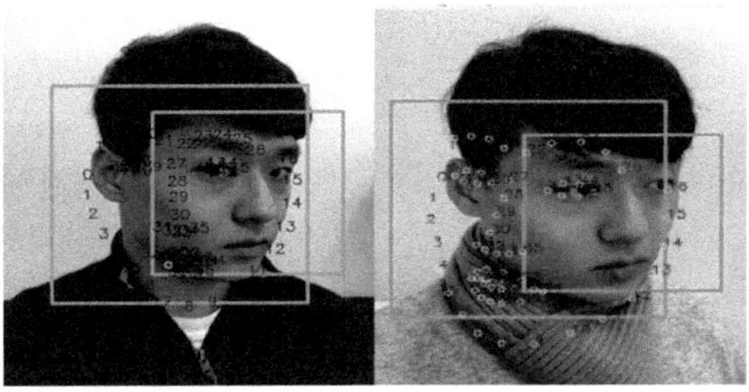

Fig. 7. Result of face detection using Haarcascade.

Figure 8 shows correctly detected one face rectangle per image using the proposed scheme of LBPH and HOG a prior trained face feature detector.

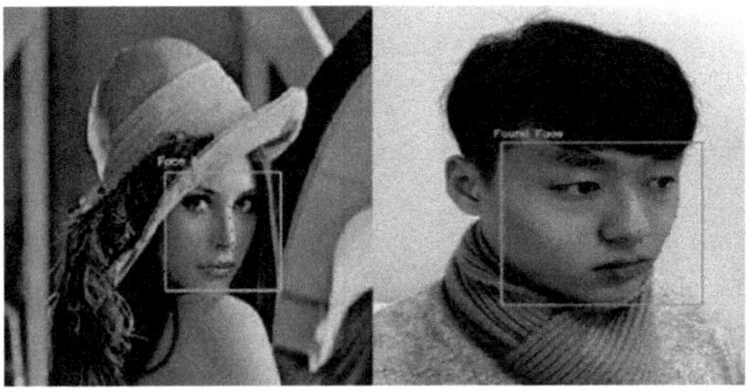

Fig. 8. Result of face detection using LBPH based HOG and SVM scheme

4 Using LBPH Based HOG and SVM Detect Faces

All previous experiments are experiments on a single image.

But this time I experimented with multiple facial images. Figure 8 shows the results of face detection for multiple face images. Experimental results were satisfactory. Multiple Face Detection Results In this paper, although two face detection results are shown, more human face detection is possible (Fig. 9).

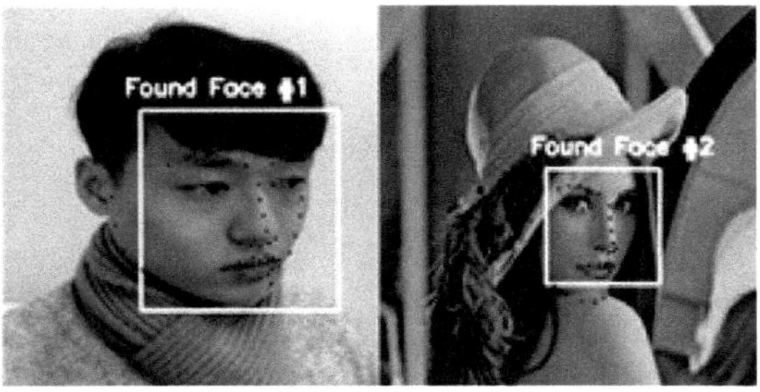

Fig. 9. Results for multiple face detection

5 Conclusion and Future Research

In this paper, an efficient face recognition scheme is proposed and implemented using LBPH, HOG, and linear SVM object detector, which is useful for rapidly spreading biometric data recognition into market. By merging LBPH algorithm into conventional Haarcascade scheme, recognition performance to detect face region is improved for not only front-view facial images but also slanted ones. In this paper, we show the results for single face image and multiple face image. The development of this study is to detect face detected by face detection and deep learning on real - time image.

Acknowledgment. This work was supported by Korea Research Foundation (NRF-2017R1D1A1B03030229).

References

1. Face Recognition Using Local Binary Patterns Histograms (LBHP) on an FPGA-Based System on Chip (SoC) (2016)
2. 300 Faces In-The-Wild Challenge: database and results (2016). Author, F., Author, S., Author, T.: Book title. 2nd edn. Publisher, Location (1999)
3. One Millisecond Face Alignment with an Ensemble of Regression Trees LNCS Homepage. http://www.springer.com/lncs. Accessed 21 Nov 2016
4. 박재호, IT Expert 임베디드 리눅스, 한빛 미디어 (2002)
5. Abdenou, H., Ahonen, T., Pietikinen, M.: Face description with local binary patterns. IEEE Trans. Pattern Anal. Mach. Intell. **28**(12), 2037–2041 (2006)

6. Ojala, P.M., Pietikäinen, M., Harwood, D.: A comparative study of texture measures with classification based on feature distributions. Pattern Recogn. **19**(3), 51–59 (1996)
7. Pietikinen, M.: Local binary patterns. Scholarpedia **5**(3), 9775 (2010)
8. Kim, Y.M., Rhee, K.H.: Face Recognition using LBPH, HOG and Linear SVM Object Detector
9. Viola, P., Jones, M.: Rapid object detection using a boosted cascade of simple features

Analysis on Signal Transmission Methods for Rapid Searching in Active SONAR Systems

Woo-Sung Son[1], YoungKwang Seo[1], Wan-Jin Kim[2], and Hyoung-Nam Kim[1(✉)]

[1] Pusan National University, Busan, Republic of Korea
hnkim@pusan.ac.kr
[2] The Agency for Defense Development, Changwon, Republic of Korea

Abstract. Pulse repetition interval (PRI) based transmission is a basic method used in radar and sonar systems to search and detect a target but it takes a long time to fully search target area because only one beam is transmitted in a PRI. In order to shorten the search time, other methods are required that can transmit signals of all beams in one PRI. There are two basic transmission methods designed by considering this requirement. One is a divided pulse length (PL) based transmission method which sequentially generates signal of each beam. The other is a multi-frequency based transmission method which generates orthogonal signals of all beams simultaneously. We analyzed the characteristics and the performance of the two transmission methods by simulations considering the underwater reverberant environment and maneuvers of a moving vehicle.

Keywords: SONAR · Signal transmission method · Rapid searching

1 Introduction

Recently, vehicles for underwater exploration have been actively developed, such as autonomous underwater vehicle (AUV) and unmanned underwater vehicle (UUV). These vehicles require some techniques that can prevent unpredictable and dangerous collisions by another underwater vehicles or aquatic animals approaching rapidly toward them [1–3]. In order to prevent these kinds of collisions, far-range sonar is required to rapidly search target area and to detect moving objects. To achieve this goal, it is important to choose an appropriate transmission signal.

Pulse repetition interval (PRI) based transmission is a basic method used in radar and sonar systems. In underwater, an active sonar uses sound wave propagating at a slow speed of 1500 m/s. The PRI based transmission generates and receives only one beam's signal in one PRI. The total search time of this method is determined by the product of one PRI and the number of beams but it is not short enough to rapidly search target area. So, other transmission methods are required to detect high-speed moving objects and to search target area faster than the PRI based transmission method.

In this paper, we analyze the characteristics and performance of the divided PL based transmission (Tr 1) and the multi-frequency based transmission (Tr 2) by exploiting time and frequency division concepts, respectively. Both transmissions

S. O. Hwang et al. (Eds.): ICGHIT 2018, LNEE 502, pp. 237–242, 2019.
https://doi.org/10.1007/978-981-13-0311-1_40

generate and receive the signals of all beams in one PRI. The analysis considers underwater reverberant environment and vehicle maneuver situation to figure out which method is more suitable for underwater detection.

This paper is organized as follows. In Sect. 2, we simply introduce the characteristics of the three signal transmission methods. Simulations are covered in Sect. 3. Finally, the conclusions are drawn in Sect. 4.

2 Signal Transmission Method for Rapid Searching the Target Area

When the maximum active detection range is hundreds of meters and each vehicle maneuvers at high speed, they may collide with each other within seconds. To prevent collision, rapid search for large area is important. First, we introduce PRI based transmission, a basic method used in radar and sonar. After that, we introduce divided PL based transmission and multi-frequency based transmission that can search the target area faster than PRI based transmission.

2.1 PRI Based Transmission

PRI based transmission generates and receives one beam in a PRI as represented in Fig. 1. If five beams from 10° to −10° are required in order to search the target area once, it takes 5 PRIs to do it. So information update period becomes 5 PRIs.

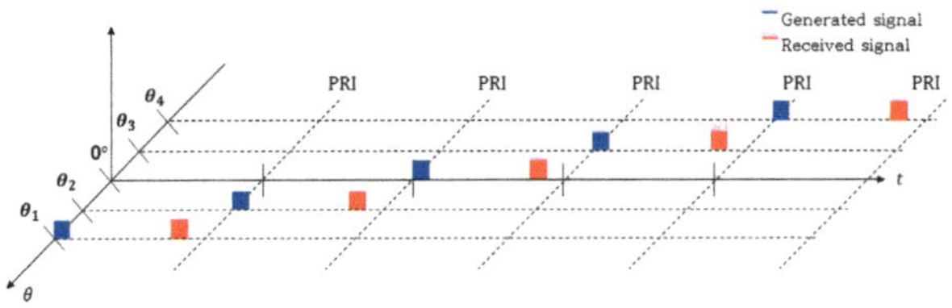

Fig. 1. PRI based transmission.

Considering distance between the vehicles (hundreds of meters) and velocity of each vehicle, information update period of 5 PRIs is too long to prevent collision caused by unpredictable maneuver situation. So the need for new transmission method is arisen that can shorten information update period.

2.2 Divided Pulse Length Based Sequential Transmission

The divided PL (pulse length) based transmission (Tr 1) is one of the basic signal transmission method that can be considered to perform rapid search for the target area and to shorten the long information update period of the PRI-based transmission. This method divides the generated signal into five beams on the time axis as presented in Fig. 2 and sequentially transmits them in a PRI to five various steering angles respectively. It can search the target area once five times faster than PRI based transmission, and the information update period is shortened to a PRI. So this is a method suitable for performing a rapid search for a target area.

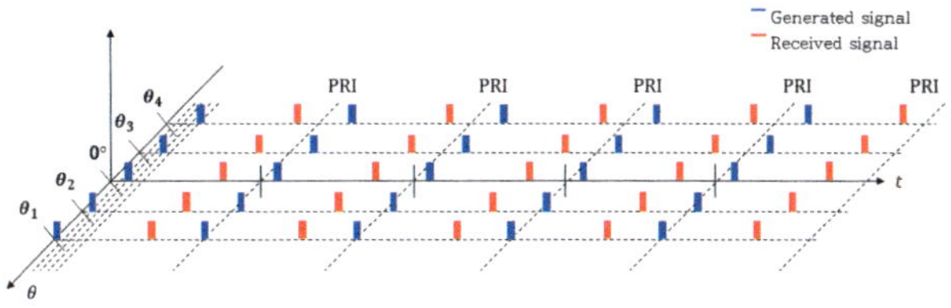

Fig. 2. Divided PL based sequential transmission.

2.3 Multi-frequency Based Simultaneous Transmission

Unlike Tr 1, the multi-frequency based transmission (Tr 2) divides the generated signal into five beams on the frequency axis as presented in Fig. 3 and simultaneously transmits them in a PRI to five various steering angles respectively. So this is also one of the basic signal transmission methods that can be considered like Tr 1.

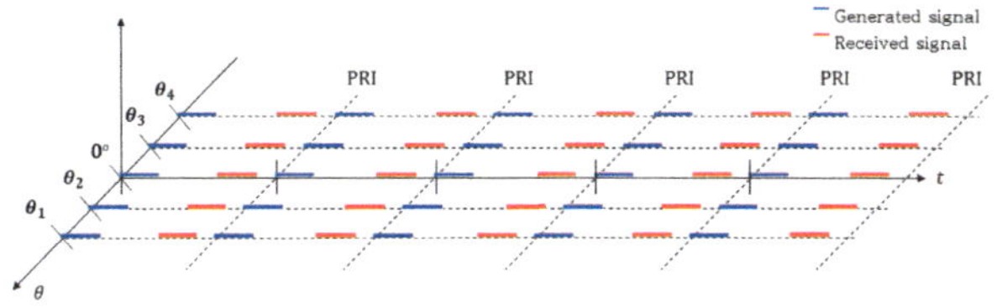

Fig. 3. Multi-frequency based simultaneous transmission.

3 Simulation

Simulations were performed to analyze the differences between Tr 1 and Tr 2 considering underwater reverberant environment and certain maneuver situation. Generated signal is divided and transmitted to five various angles ($\theta_1°, \theta_2°, 0°, \theta_3°, \theta_4°$) as presented in Figs. 1, 2 and 3. For convenience of explanation, we call a vehicle we operates, A, and an object approaching toward A, B.

3.1 Simulation 1

We set three cases of bandwidths. Case 1 is bandwidth of a single frequency signal of Tr 1. Case 2 is bandwidth of multi-frequency signals of Tr 2 which is a little narrower than that of case 1, and case 3 is bandwidth of multi frequency signals of Tr 2 that is the same as in case 1. We compared and analyzed the detection performance among three cases when the energy of each transmission beam is the same in the maneuver situation where A and B approach to each other at an angle of 0°.

In this simulation, we figured out the results that the detection probabilities are the same in the case 1, 2, 3 and the PRI based transmission regardless of transmission method or bandwidth when the energy of each transmission beam is the same as presented in Fig. 4. This result comes out because gains of adjacent beams can hardly affect center beam considering maneuver situation.

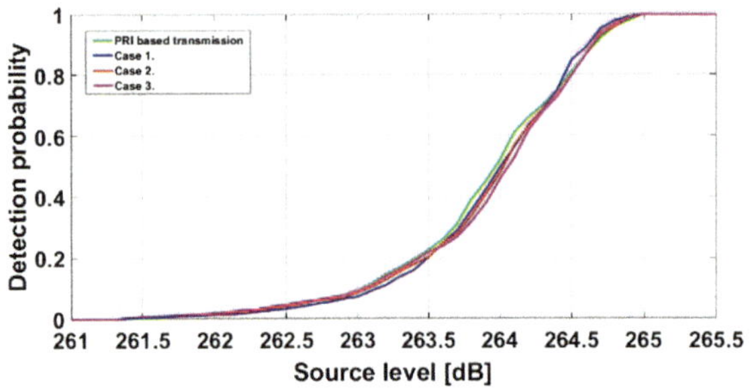

Fig. 4. Detection probabilities of PRI based transmission and three cases

3.2 Simulation 2

In simulation 1, we assumed that simulation estimates Doppler estimation accurately. In this simulation, however, we assumed that there is Doppler estimation error in the process. We compared and analyzed the detection performance between cases with different bandwidths, case 2 and 3. Figure 5 presents Doppler estimation errors (error 1, 2) in case 2, 3, respectively.

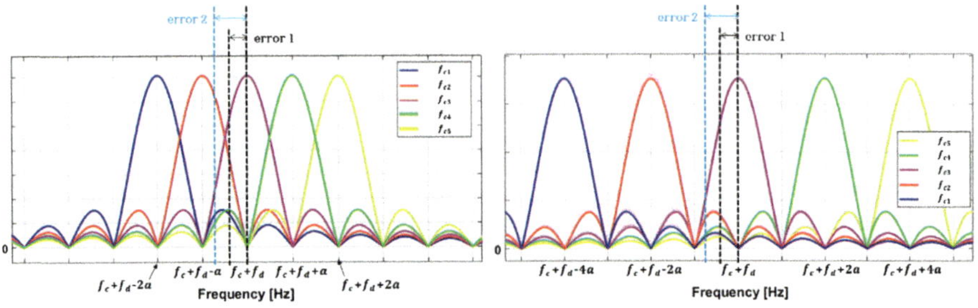

Fig. 5. Doppler estimation error 1, 2 in case 2, 3 respectively.

As a result, there is little performance difference between case 2 and 3 for error 1, 2, respectively as presented in Fig. 6. However, if we take a closer look at left figure of Fig. 5, because of an estimation error, Adjacent red color beam becomes dominant at estimated frequency. As a result, in case 2 with narrow bandwidth, there can be misunderstood that there is a target at the steering angle of the adjacent red color beam. Considering the results of simulation 1 and 2 together, even if the detection probabilities are the same, there can be misunderstood about angle of target if Doppler estimation error occurs in case that bandwidth is narrow like case 2.

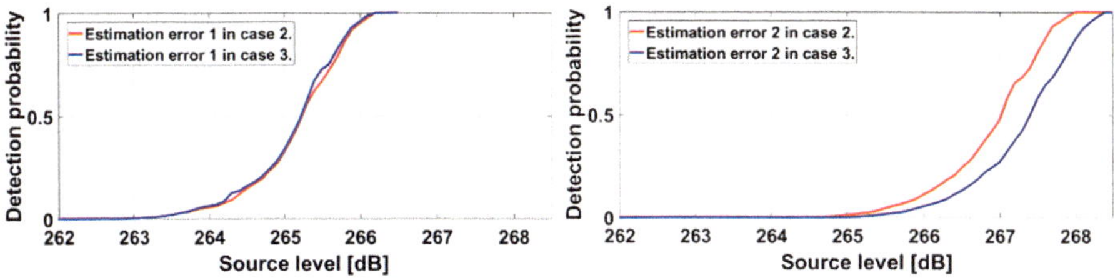

Fig. 6. Detection performance in case 2, 3 when there is Doppler estimation error 1, 2 respectively.

3.3 Simulation 3

In this simulation, we compared Tr 1 and 2 in terms of distance estimation when B approaches at 2.5° angle of A. When applying Tr 1 to the maneuver situation, B is detected in two beams sequentially as it approaches. This produces distance estimation ambiguity due to the transmission time interval between two beams. After matched filtering received signal, the signal broadens on the time axis as presented in left figure of Fig. 8, and it makes hard to estimate exact distance of B unlike Fig. 7. However, when applying Tr 2 to the maneuver situation, B is detected in two beams simultaneously as it approaches as presented in right figure of Fig. 8. In this case the distance estimation ambiguity does not occur since there is no time interval between two beams.

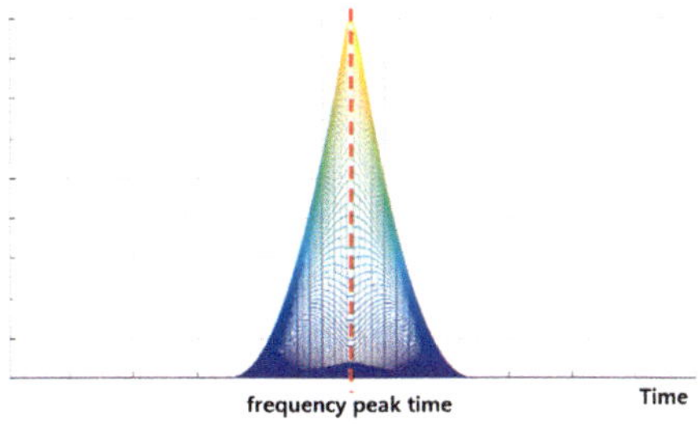

Fig. 7. After matched filtering, signal that doesn't generate distance estimation ambiguity.

As a result, Tr 2 is better than Tr 1 in distance estimation accuracy. So it is recommended to use Tr 2 when better distance estimation accuracy is needed.

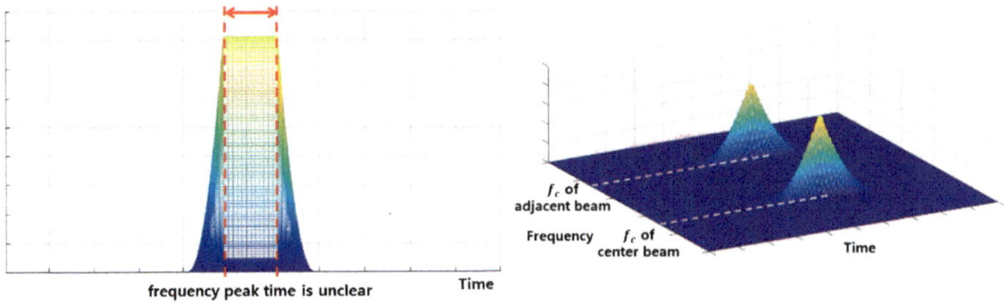

Fig. 8. After matched filtering, broadened signal (left) and signals simultaneous detected in two beams at different frequencies.

4 Conclusions

We analyzed the characteristics and the performance of divided PL based transmission and multi-frequency based transmission that can transmit signals of all beams in a PRI and search the target area rapidly.

When objects approach to each other at an angle of 0° and the energy of each transmission beam is the same, the detection probabilities are also equal regardless of transmission method. If the Doppler estimation error occurs, multi-frequency based transmission method has a problem which mis-selects the received beams.

When the target object approaches between the steering angles of the two beams, the multi-frequency based transmission is better than the divided PL based transmission in terms of distance estimation accuracy. So it may be desirable to use a multi-frequency based transmission in case that better distance estimation accuracy is needed.

Acknowledgment. This work was supported by the Agency for Defense Development, South Korea, under Grant UD150002DD.

References

1. Wynn, R.B.: Autonomous Underwater Vehicles (AUVs): their past, present and future contributions to the advancement of marine geoscience. Mar. Geol. **352**, 451–468 (2014)
2. Heidarsson, H.K.: Obstacle detection and avoidance for an autonomous surface vehicle using a profiling sonar. 2011 IEEE International Conference on Robotics and Automation Shanghai International Conference Center (2011)
3. Paull, L.: AUV navigation and localization - a review. IEEE Ocean. Eng. Soc. **38**(9), 131–149 (2014)
4. Li, Q.: Digital Sonar Design in Underwater Acoustics. Advanced Topics in Science and Technology in China. Springer, Heidelberg (2012)
5. Hodge, R.P.: Underwater Acoustics, Analysis, Design and Performance of Sonar. Wiley, Hoboken (2010)
6. Ainslie, M.A.: Principles of Sonar Performance Modeling. Springer, Heidelberg (2010)
7. Burdic, W.S.: Underwater Acoustic System Analysis. Peninsula Publishing, Los Altos (1991)

Author Index

S. O. Hwang et al. (Eds.): ICGHIT 2018, LNEE 502, pp. 243–244, 2019.
https://doi.org/10.1007/978-981-13-0311-1